Cambridge Elements ≡

Elements in the Philosophy of Biology
edited by
Grant Ramsey
KU Leuven
Michael Ruse
Florida State University

COMPARATIVE THINKING
IN BIOLOGY

Adrian Currie
University of Exeter

CAMBRIDGE
UNIVERSITY PRESS

CAMBRIDGE
UNIVERSITY PRESS

University Printing House, Cambridge CB2 8BS, United Kingdom

One Liberty Plaza, 20th Floor, New York, NY 10006, USA

477 Williamstown Road, Port Melbourne, VIC 3207, Australia

314–321, 3rd Floor, Plot 3, Splendor Forum, Jasola District Centre,
New Delhi – 110025, India

79 Anson Road, #06–04/06, Singapore 079906

Cambridge University Press is part of the University of Cambridge.

It furthers the University's mission by disseminating knowledge in the pursuit of
education, learning, and research at the highest international levels of excellence.

www.cambridge.org
Information on this title: www.cambridge.org/9781108727495
DOI: 10.1017/9781108616683

© Adrian Currie 2021

First published 2021

A catalogue record for this publication is available from the British Library.

ISBN 978-1-108-72749-5 Paperback
ISSN 2515-1126 (online)
ISSN 2515-1118 (print)

Comparative Thinking in Biology

Elements in the Philosophy of Biology

DOI: 10.1017/9781108616683
First published online: February 2021

Adrian Currie
University of Exeter

Author for correspondence: Adrian Currie, a.currie@exeter.ac.uk

Abstract: Biologists often study living systems in light of their having evolved, of their being the products of various processes of heredity, adaptation, ancestry, and so on. In their investigations, then, biologists think comparatively: they situate lineages into models of those evolutionary processes, comparing their targets with ancestral relatives and with analogous evolutionary outcomes. This Element characterizes this mode of investigation – 'comparative thinking' – and puts it to work in understanding why biological science takes the shape it does. Importantly, comparative thinking is local: what we can do with knowledge of a lineage is limited by the evolutionary processes into which it fits. In light of this analysis, the Element examines the experimental study of animal cognition and macroevolutionary investigation of the 'shape of life', demonstrating the importance of comparative thinking in understanding both the power and limitations of biological knowledge.

Keywords: comparative method, homology, Morgan's canon, contingency, convergence

ISBNs: 9781108727495 (PB), 9781108616683 (OC)
ISSNs: 2515-1126 (online), 2515-1118 (print)

Contents

Cats versus Dogs

How do biologists think? That is, how do scientists studying the living world conceive of their subject? Well, in bountiful ways, biologists are about as varied as life. There is no one way that life is, nor a single perspective best for understanding it. However, some perspectives are common, widespread, and fundamentally biological: they glom onto something that makes life 'life'. My aim in this Element is to introduce and examine one major biological perspective – what I'll call 'comparative thinking'. What is that? Let's start with cats.

It would be unnecessarily divisive (if however true) to argue that cats are better than dogs. I'll instead introduce comparative thinking by discussing how cats are special in one important respect. Let's consider some skulls, beginning with the domestic cat (Figure 1).

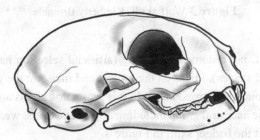

Figure 1 Cat skull, Kimberly Brumble

We'll not need fancy anatomical knowledge here. Notice the gap between the molars at the back and the incisors towards the front of the jaw. The molars continue before a short gap, after which the foreteeth begin. Let's look at another skull (Figure 2).

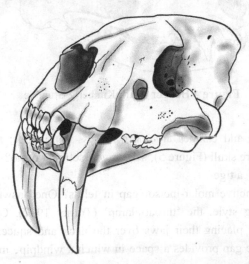

Figure 2 *Smilodon fatalis*, Kimberly Brumble

This is the extinct North American felid *Smilodon fatalis,* the saber-toothed lion. Here, the molar-incisor gap is even more pronounced. This gap is a distinctive feature of cat skulls. Consider, for instance, this representative dog skull in Figure 3.

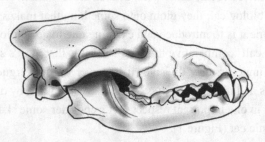

Figure 3 Wolf skull, Kimberly Brumble

This is a wolf, not a domesticated dog (artificial selection has warped dog skulls into weird shapes). Focus on the molars and incisors: the gap, insofar as there is one, is much less pronounced. In this sense, dog skulls are more typical of *Carnivora,* the ancestral group including dogs, cats, bears, weasels, and their allies. Check out the badger skull in Figure 4.

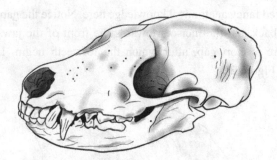

Figure 4 Badger skull, Kimberly Brumble

Again, molars and canines are more or less continuous: this is no cat. Consider one more skull (Figure 5), asking yourself: cat or no?

Yes: it's a cat – a tiger.

Why the distinctive molar-incisor gap in felids? One answer lies in their distinctive killing style, the 'throat-clamp' (Turner 1997). Cats standardly dispatch prey by placing their jaws over the neck and squeezing, blocking oxygen flow. The gap provides a space in which a windpipe may be blocked

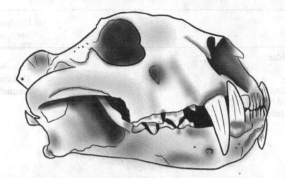

Figure 5 Tiger skull, Kimberly Brumble

or crushed. Other *Carnivora* use throat-clamps on occasion, but not to kill. Cats are built to dispatch prey using that manoeuvre.

So, cats are special insofar as they have a distinctive killing style and a morphology built to match.[1] Don't put too much weight on 'specialness': I'm sure there are similar arguments to be made about, say, dogs. Claims about uniqueness or specialness in biology are typically made against a contrast class and turn on how we describe the critters we're interested in (see Tucker 1998; Wong 2019). When we're focusing on jaws and that dental feature, cats come out as special. Perhaps if we are interested in who's the best girl, boy, or otherwise, dogs would too.

Feline specialness comes out especially in an evolutionary context, particularly if we contrast them with marsupials. Cats are placental mammals, a group with deep roots in the Jurassic period. Like other mammals, they didn't come into their own morphologically until the closing of the Mesozoic Era, after which they spread over much of the Earth. Meanwhile, the until-recent isolation of South America and Australia enabled a quite different bunch of mammals to flourish: marsupials. Over the post-dinosaur Cenozoic, groups of marsupials took to predatory ways of life, and often their skulls would evolve to resemble those of dogs. Have a look at Figure 6.

The figure maps two dimensions of skull morphology using a *principal component analysis*. Roughly, a principal component analysis takes data from a complex set of dimensions and 'summarizes' them into a smaller set while retaining patterns from the original data (see Lever, Krzywinski, & Altman 2017). So, the dimensions represent ways in which a skull's morphology might be transformed, drawn from data about variation in mammalian skulls. Across these dimensions, Goswami et al. (2009) arrange various lineages. The pictorial images represent placental mammals, the letters marsupials. Placement, then,

---

[1] For more sophisticated discussion of functional morphology, see Love 2003, 2007 and Turner 2000.

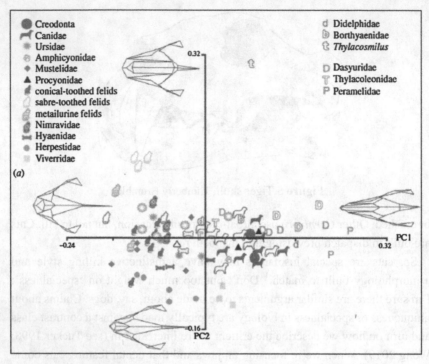

Figure 6 Principal component analysis of mammal skull shape. Detail from Goswami et al. 2011, 6 © Royal Society.

captures that lineage's skull morphology in terms of the two dimensions. Notice that dogs cluster together slightly to the right of the diagram's centre, while cats cluster to the left of the dogs. Crucially, a fair number of letters – marsupials – overlap with 'dog-space'. But none do so with 'cat-space'. Marsupial predators evolved a doglike skull at least four times, yet no known marsupial has evolved a cat skull. Why no marsupial cats?

Hypotheses concerning the uniqueness of cats' jaw morphology focus on development (see Goswami et al. 2009). Where placentals have several 'eruption points' (locations in the jaw where molars emerge), marsupials have one: molars develop one at a time on each side of the jaw, shifting along like a conveyer belt. This may mean marsupials have less scope for specialization, blocking their capacity to develop catlike dentition. Due to this (the thought goes), if you're a marsupial, dog-space is accessible but cat-space is not.

Another hypothesis identifies marsupial birth as the culprit. Marsupials are born comparatively underdeveloped, emerging from the mother's womb and climbing to the safety of her pouch. Climbing to the pouch requires a tight jaw grip, necessitating a well-developed skull. Needing a strong jaw early in development might lessen the marsupials' capacity to innovate related morphology. That is, the jaw developing earlier closes opportunities for new variation. Thus, us placentals, whose development occurs in the relative safety of the womb, might be evolutionarily freer to experiment with jaw and tooth morphology. These constraints potentially explain why our marsupial cousins evolve doglike skulls but not catlike skulls.

So, cats are special: they have a distinctive killing style, partly enabled by their placental ancestry. The evolution of their skull shapes is enabled by the developmental plasticity of placental molars, perhaps due to multiple molar eruption points or the long stay in the womb. To understand this, we examined skull morphology in cats, linking the molar-incisor gap to features of feline behaviour. We then expanded our search, comparing cats to other *Carnivora* and then to marsupials in iterative steps. These steps build what I'll call an 'evolutionary profile': we situate felines and other lineages within their developmental, phylogenetic, and adaptive contexts across multiple scales. Why so much comparison? And why those comparisons in particular? And further, why does this kind of reasoning seem so distinctively biological?

Here's a short answer. Two sets of processes shape life. First, some act as glue connecting lineages: sex, cloning, growth, and the fusion of populations create causal, ancestral, connections. Second, there are processes creating variation within and between lineages: mutation, meiosis, death, and birth – the vagaries of survival, luck, and the effects of environmental pressure. Understanding of these processes underwrites how biologists work with comparisons between organisms. Cats' unique morphology is owed both to developmental capacities and constraints bequeathed by their placental heritage and to natural selection driving them towards a predatory niche. Processes building causal connections and those generating variation and adaptation – and their integration – fundamentally underwrite comparative thinking in biology.

The long answer is the rest of this Element.

I'm going to argue that comparative thinking is a distinct biological reasoning strategy. This thinking and the methods relating to it are deeply theoretical, informed and enabled by our understanding of the two kinds of processes mentioned earlier.

Why 'comparative' thinking? Do all biological comparisons involve ancestry and adaptation? No. So, why then the label? A few reasons. First, I'm referencing a set of techniques that biologists often call 'the' comparative method. This typically involves combining phylogenetic information with methods of organizing data about variation across taxa (such as principal component analysis). When biologists talk about comparative methods, they're typically referring to techniques closely aligned with comparative thinking. Second, there's a tradition of understanding biology in terms of 'styles of thinking'. These involve a canonical explanatory schema and an accompanying perspective or point of view. As we'll see, Ernst Mayr argued that Darwinian perspectives matter because they lead us from 'essentialist thinking' to 'population thinking'. I reckon comparative thinking is another such style. But the label is imperfect: it captures a major and widespread way that biologists make comparisons, but it doesn't capture all comparisons biologists make.

This Element's structure is straightforward: Section 1 puts comparative thinking on the table; Sections 2 and 3 put it to work.

Section 1 will introduce and analyse comparative thinking. I'll begin by dipping my toe into the rather tricky conceptual territory comparative thinking bumps up against, specifically the notions of 'homology' and 'homoplasy'. I'll then consider two kinds of inferences: one, 'phylogenetic', follows ancestry; the other, 'convergent', follows adaptation. I'll go on to argue, via a discussion of adaptationism, that comparative thinking critically involves the integration of both ancestry and adaptation. This will lead me to analyse the epistemic situation of comparative thinking, emphasizing its locality: comparative inferences are restrained by phylogeny and sensitive to description. Finally, I'll discuss how comparative thinking matters for determining the significance of biological studies, discoveries, and results.

In Section 2 I'll examine how comparative thinking matters in experimental contexts by considering comparative psychology. In particular, I'll argue that a much-discussed principle, 'Morgan's canon', is made sense of via comparative thinking. First, Morgan's canon, which tells us to prefer certain kinds of hypotheses about animal cognition (namely 'simple' ones), can be understood in comparative terms: it tells us to prefer hypotheses which make sense in an adaptive, ancestral context. Second, this approach is revelatory of the experimental strategies comparative psychologists adopt. The canon can both be understood as a guide to which hypotheses to accept and to what kinds of confounding variables to control for in experiments. Understood thusly, Morgan's canon rightly places comparative thinking as central to experimental approaches to animal cognition.

In Section 3 I'll shift to more theoretical territory, considering conceptual and empirical work concerning the 'shape of life', that is, the contingency or otherwise of macroevolution. There, I'll argue that the locality of comparative thinking undermines attempts to understand life's contingency in general terms. As we'll see, both conceptual and empirical grip is lost at grand scales. I'll argue instead that 'contextual contingency theses', that is, those grounded in particular ancestral and adaptive nexuses, make for powerful and productive research.

What, then, is my overall aim? I think that comparative approaches play a profound role in the biological sciences. If you'll permit me a little pretension, I want to articulate the essence of comparative methods in biology, which I call 'comparative thinking'. To do so, I'll analyse and abstract from what I take to be particularly clear and telling examples of those practices. I'll also stick to a pretty abstract level of detail – – I'll do my best to keep things simple scientifically (and at least attempt the same philosophically).

Why should we want such a thing? For philosophers and those interested in an abstract understanding of scientific practice, the dividends are clear. Philosophers of biology have attended a fair bit to experimentation and a whole lot to models (particularly if we construe 'model' as including biological theory), but little space has been devoted to the biological tripod's third leg: the use of comparisons to investigate the living world. In examining that leg, we'll get a deeper understanding of the methods themselves; their applicability; how they play into biological and other sciences; and how this interacts with practices of integration, categorization, and the warrants underwriting biological claims. Comparative thinking, I'll suggest, is often critical both for how biological claims are established and for driving biological research.

For practising scientists the rewards are less direct. Nonetheless, I've a few things to bring to the table. My analysis of comparative thinking emphasizes the differences between biological and some non-biological sciences. Scientists are prone to worries about legitimacy: 'physics envy' at its most pernicious – the idea that biological research should in some sense follow or mock research in the physical sciences. But different kinds of systems require different research approaches and tools; there is no one way of being a successful scientist. Seeing the critical function of distinctly biological processes in shaping life suggests that the differences between biological systems and those that, say, physicists tackle are such that biologists shouldn't do science as physicists do: envy is misplaced. That is, the ontological differences between the living and non-living world necessitate different research strategies (Currie & Walsh 2018; Candea 2019). Moreover, understanding comparative thinking can advance our understanding of comparative psychology and macroevolutionary theory and, I suspect, more besides. It is worth noting that comparative thinking, although rooted in Darwinian biology, can in some circumstances extend to the social sciences and human history, so long as there are social analogues of ancestry, adaptation, and other biological processes (Currie 2016). More work is required to understand how comparative thinking plays out in human-focused science, but a firm understanding of comparative thinking in biology underwrites that further project. Finally, I think (well, at least hope!) that the kind of explanation I'll provide herein can inform biologists in a general way: why comparative thinking is so important, its role in scientific practice, and its limitations.

If you're au fait with contemporary philosophy of biology, you might at this juncture be suspicious. Not only have I led with charismatic eukaryotes – mammals even – I've also appealed to a perhaps played-out biological theme: evolution, ancestry and adaptation. Despite being well worn, I think there is more to be said about how biologists' conceptions of evolution matter for how their work proceeds. To wit, biologists often think comparatively, conceiving of biological systems as lineages shaped by ancestry and adaptation. Considering comparative thinking doesn't simply help us understand how this style of reasoning works and why, it also helps us understand when it breaks down, when it doesn't work, and why comparative thinking alone isn't sufficient for understanding all life. Comparative thinking is historically rooted in understanding vertebrate evolution, developed as it was in nineteenth-century comparative morphology and palaeontology. So, I hope you'll forgive my focus on cats, parrots, and fish.

I should say what this Element is not.

I'm not attempting to systematically document the uses of comparative methods, nor tell their history. Comparative methods themselves have a pre-Darwinian ancestry, and there is a fascinating and complex discussion to be had about how these various concepts shifted across the Darwinian divide, developing into their modern, complex, diverse forms (Brigandt 2003). The tools, research questions, and methods attached to comparative methods have been adopted in a wide range of sub-disciplines and contexts. I'll touch on some of these, but it shall fall well outside of exhaustive. This Element is not an introduction to, nor textbook regarding, comparative methods in biology. Within, you'll find very little discussion of the techniques involved in phylogenetics or the statistics underwriting principal component analysis. This will not introduce the student in a systematic way to comparative methods (for that, see Harvey & Purvis 1991; Harvey & Pagel 1991; Sanford et al. 2002; Uyenda et al. 2018).

A philosophical account of comparative thinking is critical for understanding much biology: the nature of living systems, why biologists study life as they do, and what licenses the inferences they make. Making some headway on this is my task going forwards.

1 Comparative Thinking

I've said biologists often think comparatively, situating their subjects in terms of ancestry, adaptation, and other biological processes. But what does that mean, and why does it matter? I'll start by characterizing some of the central concepts involved in comparative thinking, before highlighting two kinds of inference. Based on these inferential structures, I'll argue that paradigm comparative thinking is integrative: it is in combining ancestry and adaptation that comparative thinking is powerful. With this in place, I'll discuss the epistemic properties of comparative inferences, highlighting what I'll call their 'locality'. Finally, I'll sketch comparative thinking itself and discuss its role in driving biological research via an account of biological significance. This part is something of a crash course, so hold onto your hat!

1.1 Comparative Concepts

Two concepts matter particularly for comparative thinking: *homology* and *homoplasy*. These underwrite how biologists identify and construct kinds (Kendig 2015); they have a long – pre-Darwinian – history and have been co-opted and transported into a host of contexts, from their original homes in morphology to molecular genetics, cytology, and developmental biology. And in these new contexts the concepts take on different meanings, even as the introduction of Darwinian theory radically altered their history (Panchen 1999; Brigandt 2003).

An early definition, rooted in the comparative anatomy that was nineteenth-century palaeontology, is Richard Owen's notion of homology as 'the same organ in different animals under every variety of form and function' (Owen 1843, 379). Compare the wings of a kiwi, a penguin, and a hummingbird. These are each used for differing purposes: penguins for swimming, hummingbirds for flight, and kiwis for ... well, they're not for flying or swimming at any rate. Yet, they are nonetheless continuous: each are *wings*. They are also *forelimbs*, aligned with pectoral fins, quadrupedal forelimbs, and our arms. This continuity, or sameness of organ, is homology. What is the metaphysical nature of this sameness (Ramsey & Peterson 2012)? We might think of it in typological terms: homologies pick out 'types' – perhaps in some platonic way (although we'll meet less metaphysically weighty examples later). Perhaps the sameness is merely similarity, or a special kind of similarity, although the divergences between the kiwi, penguin, and hummingbird wings make establishing just what kind of similarity tricky (Currie 2014). There is an ongoing debate as to whether we should understand homologies in terms of types (Rieppel 2005; Brigandt 2007) or as evolutionary individuals (Ghiselin 2005, Ereshefsky 2009; Wagner 2014, chapter 7 attempts a synthesis).

On another approach, Paul Griffiths has suggested we think of homology as a *phenomenon,* that is, an explanation-demanding pattern (Griffiths 2007; Suzuki & Tanaka 2017). We recognize a continuity when we survey avian wings and identify them as forelimbs. What explains homologous phenomena? By conceiving life in evolutionary terms, we explain homology-as-phenomenon as the result of continuity in ancestry. Hummingbirds, kiwis, and penguins all have wings because they inherited them: we can trace the continuity along lines of parents and offspring.

Homologous traits may be conceived as parts of evolutionary individuals. On this view, when I say 'wing' I refer to a spatio-temporally continuous, but widely distributed, individual, similar to how I might refer to myself at earlier and later times, or to a set of fast food franchises as being 'the same' restaurant. In addition to suggesting an answer to our metaphysical question concerning sameness, it also suggests an answer to our conceptual quandary regarding homology's definition: two traits are homologous just when their similarity is explained by their being inherited from a common ancestor. Or, if we dislike the explanatory spin, we could more traditionally say:

> Two traits in two lineages are homologous just when that trait is also held by the most recent common ancestor of those lineages.

This is a *taxic* account of homology (Brigandt & Griffiths 2007; Ramsey & Peterson 2012; Currie 2014). It conceives of homology in phylogenetic terms. Hummingbirds, kiwis, and penguins have wings – as a homologous category – because they share a common ancestor (likely a small theropod dinosaur) that also had wings. This also gives us a definition for our counterpoint: *homoplasy*. Birds haven't a monopoly on wings: consider bats, pterosaurs, various flying insects. And these are not commonly inherited. Here, continuity is not due to their being the same regardless of function but *because* of function. Bats, pterosaurs, and birds have come to similar solutions to a similar challenge: airborne locomotion. We might say that two traits are homoplastic just when their similarity is explained by their being adapted to similar environments. Or, again avoiding the explanatory spin:

> Two similar traits in two lineages are homoplastic just when that trait is not present in the lineages' common ancestor.

So, bats' and birds' wings are homoplastic because their most recent common ancestor – some kind of early Permian reptile – was not winged. Note that the explanatory and merely phylogenetic definitions of homoplasy come apart. Just because a trait is not held by a common ancestor doesn't mean that the explanation of its presence is being adapted to similar environments, although it is a tempting guess. And the same goes for homology: if wings were present in the common ancestor of hummingbirds and vultures but were subsequently lost and then re-evolved in both lineages, then we have a homology by the phylogenetic definition but not by the explanatory definition.

The decoupling of *taxic* definitions of homology and homoplasy from explanations of their presence leads some authors – Brian Hall most prominently – to consider homology and homoplasy as continuous (Hall 2003, 2007). After all, in explaining the wings of penguins, kiwis, and hummingbirds, I combine both homoplastic (the penguins' adaptations to swimming) and homologous (the basic wing structure) aspects; and in explaining pterosaur, bat, and bird wings, I appeal to both their being shaped by similar selection pressures and their deep continuity as vertebrate limbs. Although such mingling of ancestry and adaptation is paradigmatic comparative thinking, I think it is worth resisting collapsing explanatory continuity with conceptual continuity; that is, we shouldn't run the notions of homology and homoplasy together. The basic reason, which I'll come to in the next subsection, is the epistemic difference between convergent and phylogenetic inferences. Taking the two as conceptually continuous disrespects the epistemology (Currie 2014).

So, things seem hunky-dory: we can explain homology qua phenomenon by recognizing its ancestral nature as captured by the *taxic* definition and explain homoplasy when it arises by appeal to natural selection. However, there are two basic problems. The first is that there are *non-taxic* definitions jostling for attention. The second is that notions of homology and homoplasy have, in the course of twentieth-century biology, been transported and adopted into new biological contexts; in these contexts taxic homology looks significantly less attractive.

Two main competing conceptions of homology appeal to different directions biological research has taken. *Transformative* accounts of homology are in many ways akin to pre-Darwinian conceptions but are associated with some explanations in modern evolutionary developmental biology. As Brett Calcott has explained, a paradigmatic explanation in this field will account for how a particular trait evolved by positing a set of incremental developmental steps, often whilst holding fixed some selective pressure (Calcott 2009). A classic example is the evolution of the lens-eye characteristic of cephalopods and vertebrates. From a single light-sensitive spot, small changes selected for acuity can incrementally lead to the development of an eye. That is, there is a *transformative sequence* from the light-sensitive spot to the lens eye. By transformative accounts of homology, two traits in two lineages are homologous just in case there is such a transformative sequence between them.

Developmental accounts link homology to common developmental mechanisms (Wagner 2018). Homologies between body parts (say, mother cat's tooth-gap and her daughter's tooth-gap) are inherited via a different mechanism than molecular homologies. The latter are copied via gene replication. The two strands of the double helix split apart, each half then acting as a template for the construction of new double helixes. Over time, and across generations, these form lineages linked via such replication events. By contrast, in body-part inheritance there is no mechanism using the mother's tooth morphology as a template for her offspring. Rather, shared developmental resources are passed along generations and new jaws (and cats!) grow on the basis of those resources. At base, developmental accounts of homology are geared towards capturing homologies between body parts: if the traits develop based on the same inherited resources, then they are homologous. One upshot of developmental accounts, which *taxic* accounts drop the ball on, are 'serial' homologues. Consider your own digits, or your left and right hands. Presumably, these are 'the same' organ in just the same way as your and my hands are 'the same' organs. Yet *taxic* conceptions miss these as there is no common ancestor involved. Furthermore, developmental accounts allow us to conceive of homologies as units of evolution (Brigandt 2007); it is developmental homologues that evolution acts upon and shapes over time.

As I'm interested in how these concepts play into comparative thinking, here is not the place to get into the conceptual and metaphysical minutiae. However, I am fairly optimistic that a more-or-less unified account of the concepts can be had. To see the shape of the proposal, let's ask a different question regarding homology and homoplasy: should we be monists or pluralists about the concepts?

Conceptual monists and pluralists disagree about whether, for some category, there is a single concept able to accommodate the various ways in which that category is drawn across the contexts in which it is deployed (e.g., Ereshefsky 1998; Currie & Killin 2016). That is, is there a single notion of 'homology' able to do the work we need it to in morphology, development, molecular biology, and so forth? We can think of concepts as functions taking us from sets of objects to sets of categories. Taxic accounts of homology take a set of traits, categorizing as homologous those where the common ancestor of the trait's owners also possessed that trait. Different accounts of homology lead to different divisions. For instance, because traits over evolutionary time can become decoupled from the developmental mechanisms that underwrite them (Hall 2003), some traits can count as homologues in taxic terms, but not in developmental terms. Some newts, for instance, develop digits via addition – each 'finger' is grown separately – while all other vertebrates grow them via accretion: first a paddle is grown, before the webbing between the fingers is removed. In this case, we have different developmental mechanisms (so not homologous on developmental grounds), yet these newts have relatives sharing common ancestors with digits (so homologous according to taxic conceptions).

So, various accounts of homology appear to handle some, but not all, of the cases we're interested in. Monists will insist that they can develop one account to rule them all: attempting to design concepts which manage – or near enough – to do all the work that is required (Ramsey & Peterson 2012; Currie 2014). Pluralists argue that 'homology' breaks down into several non-equivalent concepts, each required for some different purpose (Brigandt & Griffiths 2007). Aaron Novick (2018) combines a kind of monism with the flexibility of pluralism, which I think aligns nicely with comparative thinking.

On Novick's view, there is a thin notion of homology which various accounts fill out towards fulfilling varying purposes. He concentrates on taxic and developmental accounts of homology, viewing them as complementary aspects of a more general category. The abstract notion – which is general insofar as it applies to all taxa – is derived from two formal notions: 'descent' and 'character'. As Novick puts it:

> Anything can be a 'character' (i.e. used in phylogenetic systematics), provided that it yields transformation series. Likewise a phylogenetic descent relationship is simply any relationship that gives rise to phylogenetic patterns recoverable by phylogenetic analysis (Novick 2018, 8).

This takes a little unpacking. Consider the inference from catlike tooth morphology to being a cat. This requires our recognising the molar-incisor gap across cats as a *character* able to take a number of *states*. Most basically, we could distinguish between having the gap and not having it. In cats, jaw morphology takes the molar-incisor gap state, while in other *Carnivora* (badgers, for instance) jaw morphology is not in that state. We can conceive of a transformation series between these states. That is, we could construct a series of changes taking us from a standard *Carnivora* jaw with continuous molars and incisors to a jaw with the molar-incisor gap. This is what it takes to be a character. Now, when inferring along homologies, we conceive of the relevant lineages as ancestrally united by a common ancestor. That is a phylogenetic pattern, as is the distribution of character states within it (for instance, all felines possessing a molar-incisor gap). So, we can understand homology as involving two genealogical aspects: first, there being a feature that takes a number of states; second, those states are capable of forming the basis of a phylogenetic pattern. Another way of getting at Novick's idea, I think, is to say that something is a homology in the abstract case just in case it can be used in a phylogenetic inference of the sort I'll describe in the next section.

It is critical to Novick's account to see that the formal notions of descent and character are independent of the various mechanisms by which the relationships of ancestry and transformation are instantiated. Although I've been speaking of phylogenies uncarefully, with the notion of ancestry built in, phylogenies themselves can be understood as applying a formal structure to a set of character states. 'There is no principled reason why ancestors and descendants, *qua* ancestors and descendants, cannot be radically different. Phylogeny alone places no limits on the extent of possible divergence between them' (ibid, 6). A phylogeny without a mechanism of descent – in virtue of what do offspring resemble their parents? – does not explain the patterns of variation we see. And because different kinds of characters (genes, or body types, and so forth) are inherited in profoundly different ways, how we carve out those characters, what homologies there are, depend on what kind of inheritance system we're dealing with. Recall our brief discussion of genes, which form genealogies due to replication, and body parts, which form genealogies due to inherited developmental resources. Although these can both be understood as characters with relationships of descent, the mechanisms of inheritance differ. For Novick, then, to be applicable the general account of homology needs to be *enriched* with specific detail of the relevant inheritance mechanism. Because different mechanisms generate different patterns, there is no one enriched homology applicable across life.

In Novick's account, then, we can incorporate the diversity of homology concepts while still understanding their general unity. We can see how quite different-looking inferences such as, say, the inference between possessing molar-incisor gaps and being a cat, and the inference between a particular genetic complex to a similar complex in a related group, are nonetheless part of the same general inferential strategy. Another important aspect of Novick's account, which we'll return to below, is *locality*: 'the manner in which the genealogical account is to be completed will depend on the type of character in question – thence the need for local, not global, enrichment' (ibid, 6). I'll soon argue that locality is a crucial aspect of comparative thinking.

Can Novick's strategy be attempted for homoplasy as well? In principle we could say that homoplasy occurs when two lineages share character states without sharing a common ancestor with that state. But simply identifying shared character without ancestral connections doesn't explain their being shared. Further, depending on what you include, some homoplasies, particularly those across extremely distant ancestry, might not count as characters at all. So, this general notion of homoplasy needs to be enriched with mechanisms explaining their similarity. The basic convergent inference we'll focus on in the next section uses a rather general mechanism: the thought that natural selection will drive lineages towards similar solutions to the challenges raised by similar environments. I suspect that other ways of conceiving of natural selection might yield different ideas about homoplasy. And indeed, we needn't appeal to natural selection to explain the phenomenon. It could be that, for instance, similarities are simply due to chance. Or it could be that there are other, non-selective, processes that tend to drive various lineages into similar morphologies.

Regardless, as we can understand Novick's notion of homology as closely aligned to one of the inferences I'll describe it in the next section: a feature is a homology when it can be conceived as a character and employed in a phylogenetic schema, and we can understand homoplasy in terms of what I'll call a 'convergent inference'. That is, something is homoplastic just in case it can be used to inform a model of a similarity-generating mechanism such as natural selection. This basic idea requires more development than I've space here, but regardless these two very general notions, homology and homoplasy, sit at the heart of comparative thinking.

1.2 Two Kinds of Inference

With homology and homoplasy under our belts, I want to highlight two kinds of inferences we saw in our examination of feline dentition; first concerns homoplasies driven by natural selection as a mechanism, and second homology driven by ancestry.

We saw that the distinctive gap between molars and incisors in the feline skull was an adaptation for a particular killing method: cats kill via strangulation, and the gap accommodates the victim's neck. A trait is an *adaptation* when it has been selected for in the past; when it has been responsible for successful ancestors having successful descendants. To argue feline dentition is an adaptation for strangulation, then, I need to make two claims. First, a claim about function, what cat dentition is good for. Second, a claim about the trait's history: it spreads throughout feline populations as a result of fulfilling that function. The former claim is logically independent of the latter, but the latter depends on the former. Cat dentition may be good for strangulation, but may not have spread through feline populations because of that function. The former claim is ahistorical, concerning whether a certain trait is optimal (or at least efficient, or at least good) at performing some task. That is, I needn't say anything about the actual history of the trait in question, nor the history of the environments it excels in. I can make claims about what a trait would be optimal for without the trait ever having in fact been present. The latter claim, that efficiency for strangulation explains the molar-incisor gap in cats, is often supported via *reverse engineering*: because the trait is used for this purpose across the feline clade, it likely arose in light of evolutionary pressures for that particular purpose. One way of supporting this is via what I'll call a *convergent inference* (Sayers & Lovejoy 2008; Currie 2012, 2013; Vaesen 2014). With evidence of a trait's optimality, we search for examples of similar traits evolving in similar environments: that is, we seek homoplasies. This provides our abstract adaptive model empirical traction.

In addition to making claims about the optimality and fitness of feline dentition, I made a claim about ancestry: homology. Felines have distinctive dentition, and they share this in part because the trait was inherited throughout the group. Paleontologists discovering extinct feline species would most likely reconstruct them with this trait – even if the skull was not present or was heavily degraded. Here, we have two historical inferential steps. We first infer from the presence of the trait in contemporary critters to the trait's presence in their common ancestor. The presence of the dental gap across the feline clade suggests its presence in some past ur-feline. We then infer from this common ancestor to other members of that ancestral group. We might, for instance, identify remains as being feline on the basis of their having that trait. I'll call this a *phylogenetic inference* (see Sober 1991; Currie 2014, 2016; Levy & Currie 2015).

We have, then, two kinds of inference. The first takes us from a trait's suitability, optimality, or efficiency in performing some function in an environment to the trait sometimes evolving in that environment to perform that function. The second takes us from the trait's distribution in taxa to its inherited presence in hitherto unexamined examples of that group. Let's more carefully characterize these.

Phylogenetic inferences carry us along lines of ancestry. At base, a critter possessing a trait is evidence that its relatives also possess that trait. Figure 7 captures the basic idea. Here is an abstract formulation:

> A *phylogenetic inference* infers from one lineage (*A*)'s possession of a trait to another lineage (*B*)'s possession of a trait on the basis that *A* and *B* are phylogenetically related, and the trait is likely to have been inherited from *A* and *B*'s common ancestor; *Or* infers from an unknown lineage's possessing a set of diagnostic traits to it being a member of some phylogenetic group on the basis of those diagnostic traits being commonly inherited.

This is a disjunctive[2] definition of phylogenetic inference. The first disjunct infers from phylogenetic membership to a trait, the second from trait to phylogenetic membership. An inference from an unknown lineage having a catlike skull to it being a cat is of the second type, an inference from a lineage being a cat to it having a catlike skull is an example of the first type. The two inferences are unified by the factors licencing the inference. Namely, the fidelity

---

[2] An inference is *disjunctive* when the inference's soundness requires at least one of the premises to be true.

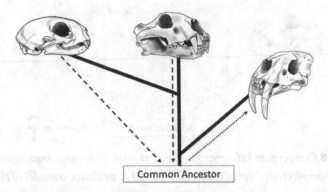

Common Ancestor

Figure 7 Phlogenetic Inference. *A two-step inference. In the first, traits of closely related lineages form the basis of an inference to the traits of the common ancestor. In the second, that trait is projected onto another phylogenetically connected lineage.* Hard lines represent ancestral relations, dotted lines directions of inference.

of the phylogenetic signal. A phylogenetic signal is high-fidelity when we have good reason to expect similarities in traits across organisms to be due to their being commonly inherited, that is, we should expect homology. When biological properties are likely to remain stable over evolutionary time, then it is relatively safe to infer from similarity in traits to common ancestry, or vice-versa. If traits are highly labile, then it is likely that similarities may be homoplastic.

The second kind of inference is via adaptation, not ancestry. Here, we infer from a common environment to common traits, or common traits to a common environment. I'll call these *convergent inferences* (Figure 8):

A *convergent inference* infers from a trait-environment match in one (or a set of) lineage(s), to the same trait-environment match in another lineage on the basis of either (1) evidence that the lineage is in a similar environment or (2) evidence the lineage possesses that trait.

Convergent inferences infer on the basis of natural selection's capacity to generate regularities. I envision these in terms of *adaptive models*. Adaptive models represent a trait as being an adaptation in light of a particular environment or niche. The model says that some trait is optimal in some environment and thus, under some conditions, one is a signal of the other. The cat's killing method is presented as a solution to a particular kind of problem: how to quickly and relatively safely dispatch high-value but dangerous prey. Strangulation, and the dental morphology enabling it, is then an adaptation to a megacarnivorous niche. The thought behind adaptive models is that natural selection leads to common solutions to environmental challenges across taxa. One way of evidencing adaptive models appeals to fit or

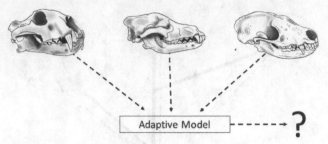

Figure 8 Convergent Inference. *In the first step, trait/environment matches across a paraphyletic,*[3] *set of lineages are used to evidence a model of that trait/ environment match. In the second step, we infer the presence of a trait or environment in another case.* Dotted lines represent inferences.

optimality: features of cat morphology are well-suited to the task, so, if such traits are available and the environment favours that task, we should expect it to evolve. This grants *prima facie* plausibility to the latter having evolved in response to selection pressure from the former. Another source of evidence, and our focus here, is *convergences*. If similar environments correlate with similar traits, this is *prima-facie* evidence for the environment having shaped the trait. Here, the strength of the inference turns on the strength of the trait-environment match.

Comparative inferences do not involve simple enumerative induction,[4] but are highly theoretical, requiring substantial commitments about the nature of the evolutionary forces shaping the traits and lineages in question. For phylogenetic inferences, we must commit to the role of ancestry in maintaining high-fidelity information across lineages; for convergent inferences, we must commit to the power of natural selection to guide lineages to similar solutions. Inferring from cat skull to cat-hood involves explaining the presence of the cat skull in terms of its bearer being a cat. Similarly, inferring from possessing a catlike skull to megacarnivory is very close to explaining the presence of the trait in question in terms of it having been selected for in that environment in the past. In this sense the two kinds of inferences – expressed abstractly – are what Angela Potochnik's has called *explanatorily independent* (Potochnik 2010).

Two explanations of a phenomenon are *independent* when each are satisfactory despite black-boxing details which the other emphasizes. Potentially, that cats' jaw morphology is an adaptation to their killing style is a sufficient explanation for their having the dental gap and it not being present in other

---

[3] *Paraphyletic* traits are descended from a common evolutionary ancestor or ancestral group but are not included in all of the descendant groups.

[4] The support of an *enumerative induction* is derived purely from the number of supporting instances.

members of *Carnivora* who didn't adapt to that killing style, without us mentioning anything about the trait's inheritance *per se*. However, the trait's inheritance and taxic distribution are critical for us establishing the epistemic credentials of the explanation: they are in Potochnik's terms *epistemically interdependent*. To put it coarsely, the explanation might be satisfactory, but we don't know if it is true without examining ancestry. For instance, we might discover that early cats had the dental gap but did not kill via strangulation. In this case, although the first part of the inference underwriting the adaptationist claim – that the dental gap is an efficient way of killing via strangulation – is true, the second part – that the trait spread throughout ancestral populations due to its efficiency in killing – is false. The thought is that an adaptationist explanation black-boxes ancestry: drawing apart the relevant contrasts doesn't require information about ancestry. However, information about ancestry is required for the explanation's truth: ancestry is black-boxed but not irrelevant for determining the explanation's epistemic status.

Convergent models, or so it seems, black-box ancestry; phylogenetic inferences, or so it seems, black-box adaptation. However, how things go within the black-box matters for the inferences' power – that is, they are epistemically interdependent.

Ancestral relationships matter for the applicability of adaptationist models because the regularities such models capture can be constrained by development. As we saw earlier, no marsupials have converged on the distinctive catlike skull, and the two going explanations for this appeal to developmental constraint. It doesn't follow that there wouldn't be selection for cat-like skulls were they developmentally available to marsupials. Conversely, natural selection can matter for phylogenetic inferences. This is because ancestral continuity can involve maintenance selection – here, high-fidelity continuity is not due to highly constraining paths of inheritance but due to major fitness costs to diverging from that trait. And because high selection might overcome developmental constraints – if some very unlikely but very fit trait were to emerge – inferences along inheritance at least to some extent involve a commitment to strong selection not acting in another direction. So, in practice, the two inference-types we've examined, despite their apparent explanatory independence, often involve heavy commitments regarding a lack of interference from ancestry (for convergent inferences) or adaptation (for ancestral inferences) (Griffiths 1996; Sansom 2003; Pearce 2011).

1.3 Adaptationism and Integration

In this section I'll show how the integration of ancestry, adaptation, and other evolutionary processes are the centrepiece of comparative thinking. That is, paradigm comparative thinking draws together both phylogenetic and convergent inferences to construct what I'll call an 'evolutionary profile'. To do so, I'll put pressure on pluralist defences of adaptationism, such as that suggested by the notion of 'explanatory independence' earlier. Convergent inferences are closely aligned with adaptationism: a focus on the explanation of biological form via biological function, in particular, the power of natural selection to shape lineages to particular niches. Since 1970s this approach has met with many – often highly rhetorical – challenges.

The most famous challenge to adaptationism is Richard Lewontin and Stephen Jay Gould's *Spandrels of San Marco* (Gould & Lewontin 1979; Forber 2009). The paper, in effect, provides three interrelated objections to adaptationism. First, adaptationists ignore alternative hypotheses for why biological traits are the way they are. Architectural or developmental constraints could also account for robust patterns across organismic traits. Second, adaptationists categorise, 'carve up', their organisms via adaptations. That is, the traits – the things to be explained – are understood as adaptations, effectively closing off other ways organismic parts (or wholes!) might be conceived. Third, as a result of ignoring non-adaptationist possibilities, adaptationists generate ad-hoc 'just-so stories' in light of apparently falsifying evidence. Gould and Lewontin pitch adaptationist explanations as in conflict with more sophisticated developmental-based explanations, or more sophisticated ways of thinking about natural selection (as including path-dependent exaptations,[5] or – although not mentioned in the original paper – niche construction where lineages shape their environments as well as *vice-versa*)[6].

Philosophical defences of adaptationism have often involved distinguishing between various kinds of adaptationist commitments, or identifying a kind of explanation that adaptationist explanations are particularly good at providing; thus adopting a pluralist strategy. For instance, Peter Godfrey-Smith (2001) distinguished between empirical, explanatory, and methodological versions of the idea (and Tim Lewens expands these further, 2009). While an empirical adaptationist is committed to life in fact being shaped by natural selection, explanatory adaptationists take those parts of life which happen to have been shaped by natural selection to be the things most deserving of explanation, and methodological adaptationists think that adopting an adaptationist approach is

---

[5] An *exaptation* evolved for some other purpose but was later co-opted for a new one.

[6] This clash over adaptationist versus non-adaptationist modes of biology is often characterized in terms of a battle between an old guard of adaptationist 'neo-Darwinian' folks with roots in the Modern Synthesis of the 1940s and '50s, and a new-guard of Extended Synthesis folks who aim to expand the biologist's theoretical toolkit (Pigliucci & Müller 2010). This has been characterized as everything from an overblown teacup storm to a dramatic Kuhnian paradigm shift.

a fruitful way of doing biology. Methodological adaptationists, then, deny that life is largely shaped by adaptation and that adaptation is explanatorily privileged: rather, if we start with adaptationist questions, we'll build our way towards a rich understanding of life.

Sober and Orzack (1994), by contrast, associate adaptationism with optimality-modelling. For them, optimality models identify certain phenotypes as being evolutionarily stable in a technical sense, that is, any other available phenotype would be lower in fitness.[7] For Orzack and Sober optimality models are 'censored' models – they involve abstraction from factors such as drift and developmental constraint. Such a model can show that natural selection is a sufficient explanation for a trait when, so they argue, it is statistically accurate. Orzack and Sober's strategy is to try and show how adaptationism involves testable hypotheses through underwriting optimality models.

We can, then, identify a pluralist strategy in defending adaptationism: it is an autonomous mode of explanation which is not in competition with others ways of doing biology. This works fine so far as it goes, as do the two isolated inferential structures I earlier labelled 'phylogenetic' and 'convergent'. But in comparative practice these modes of explanation and inferences are not kept separate but intimately interwoven (c.f. Brigandt 2010; Brigandt 2013). I'm going to make this point in two ways, first by considering how they play out in our earlier case study, second by considering the notion of 'parallel evolution'.

We've seen how cats are special because of their molar-incisor gaps. We accounted for this specialness in several ways. There was a distinctive killing manoeuvre, the throat-clamp, which the gap plausibly aids: it is an adaptation to their particular killing style. One could on this basis go hunting for other animals who kill via strangulation and see whether they have similar adaptations, thus forming a convergent inference. However, if you go looking amongst the marsupials, you'll be disappointed. And this raises another sense of specialness. Cats are able to have the jaw morphology they do in virtue of the flexibility of the developmental systems bequeathed from their placental ancestry. For whatever reason (we surveyed a few options) marsupials are potentially unable to evolve in that direction. Here, we're thinking both of lines of ancestry and of how evolutionary paths are constrained.

Following the pluralist strategy, you might insist that the explanation of cat dentition in terms of strangle-hold optimality and the explanation of marsupials not occupying cat-skull-space in terms of developmental constraint, are strictly speaking independent. And indeed we can abstract them as such. I just did so.

---

[7] This technical notion is not the only way we might understand optimality models: we can also understand them in terms of being (something like) 'well engineered'.

However, this doesn't capture in practice how these investigations motivate and feed into one another, nor does it capture the importance of their integration. Shifting from a focus on abstract inferences and isolated explanations, what I think we see in the mammal-skull-space from the introduction is the development of an *evolutionary profile* for the lineages in question. This profile involves the construction of an integrative model: one which combines both ideas about selective pressures and how particular taxa – with the various developmental constraints and resources particular to them – respond to those pressures. Examining both the spread and the peculiarities of traits across placental mammals, putting forward and testing hypotheses regarding their adaptation, comparing these with other related groups and then considering how the difference might be explained in terms of differences in their developmental resources, builds a picture of the evolvability[8] of that lineage and provides a ground from which to build and test narratives about the evolutionary paths the lineage followed. While we might split the inferences and explanations involved apart, what makes comparative thinking so powerful is its being geared towards building an integrated evolutionary profile.

The importance of integration is seen in another conceptual ripple, which some think challenges the apparently discrete notions of homology and homoplasy we dissected in the previous section. In addition to the notion of *convergence*, which is often conceptualized as two different lineages evolving similar traits from different starting points, we can also imagine two lineages evolving similar traits *in parallel* (Hall 2012; Pearce 2012; Powell 2012; Ghiselin 2016). Compare two hypothetical scenarios. In the first, imagine, some marsupial overcomes the constraints of their development (their teeth begin erupting differently, say) and then evolve an incisor-molar gap for the purpose of killing via throat-clamps. In the second, let's imagine that the common ancestor of cats didn't have the incisor-molar gap, and that the various cat lineages then proceeded to evolve the trait along similar lines to each other. On a taxic conception, we'd agree that both of these are homoplastic: neither the common ancestor of the placentals and the marsupials nor the common ancestor of the cats had the incisor-molar gap. And both accommodate adaptationist readings: the gap evolved due to its role in throat-clamps. Yet the two, we might insist, differ. In the latter case, we learn something about cats, while in the former we learn something about the relationship between molar-incisor gaps and throat clamps. The parallel evolution of cat skulls has a narrower taxic range than convergence. We might then be tempted, in the parallel case, to say that for all

---

[8] 'Evolvability' is a tricky concept, at a first pass we could understand it as the capacity of a lineage to evolve various traits (see Brown 2014).

intents and purposes this is a homology rather than a homoplasy. After all, it seems the continuity in developmental resources between the common ancestor and the cats is doing a lot of work in explaining the evolution of the trait. And indeed, one could take the parallelism as evidence of cat-hood just as one could take proper taxic homologies as such.

One lesson we might to take from parallelisms is to say that the distinction between homology and homoplasy is graded; another is to say that parallelisms are in fact a kind of homology (Ramsey & Peterson 2012). Hall (2007) takes something like the graded route, arguing that any convergence involves shared developmental resources which potentially go some way towards explaining those convergences. I think Hall's point about the epistemology of convergence is right, but don't think the conceptual part follows (for reasons we needn't get into, Currie 2012). The relevant point here is this: in explaining convergence or parallelism, combinations of ancestry and adaptation are bought together. Again we see the integrative hallmark of comparative thinking.

1.4 Locality

I've characterized two kinds of inference and explanation – ancestral/phylogenetic and adaptive/convergent – and demonstrated how these are integrated. This integration is central to comparative thinking. The aim of this section is to lay out some features of what I'm calling evolutionary profiles and discuss their epistemic power and limitations.

'Comparative thinking' doesn't involve any old comparison between two biological features. It involves a comparison between at least two biological features which are *situated in an evolutionary context*, where an 'evolutionary context' is the combination of relevant evolutionary factors: ancestry, adaptation, environment, and so on. The comparison between the dental organization of cats and dogs is motivated by considerations of their different adaptations in the context of their both being *Carnivora*. The various lifeways and body-plans of *Carnivora* make a difference to both the kinds of comparative inferences which are licenced and what our research questions are. In short, the significance of the similarities and differences, and what inferences and explanations may be drawn from them, depend critically on the ancestry and adaptation of the cases.

I'm going to argue that comparative evidence is *local*, but what do I mean by 'locality'? What can be done with comparative data is fundamentally constrained by the local conditions of its production: the measurements actually made on the specific specimen, the data's provenance, and so forth. This is true, but I think true of evidential reasoning in general (Leonelli 2016; Boyd 2018;

Currie 2019a). By 'local' I mean that the details of ancestry and adaptation fundamentally constrain the kinds of inferences that can be drawn. The hypotheses that can be tested are at base limited in scope. In Section 3 I'll argue that this constrains the utility of comparative evidence for informing general questions about 'the shape of life'. I'll cash out this locality in terms of three features: comparative methods are *situated, specific,* and *sensitive.*

Comparative methods are *situated* within a phylogenetic context. The analyses of cat skull morphology are founded on comparisons between placental and marsupial mammals. This situating leads us to certain questions and hypotheses: what is different about those clades such that one, and not the other, evolves catlike skulls? This situating also restricts the scope of the inferences at hand. Models of dog skull evolution are restricted to those clades with the requisite skull design. The model may be paraphyletic, including some sets of marsupials and some sets of placentals, but not all, but nonetheless without further study the applicability of the model is restricted to those groups. The evolution of cat skulls has been linked to the emergence of the press-and-hold bite (Van Valkenburgh & Jenkins 2002). This is a precondition for the throat-clamp killing style cats employ. However, without further information it would be a mistake to infer the same in different phylogenetic contexts, say, the marsupials.

Comparative inferences are *sensitive* to description. A property is sensitive to description just when that property's holding depends upon how it is described. For instance, my accent is somewhat unusual for a lecturer at the University of Exeter – which I am – but it is not unusual for a New Zealand citizen – which I also am. So, whether my accent is unusual or not depends upon the description I'm falling under. Similarly, there are many senses in which cat skulls are not unique: as members of *Carnivora,* they sport the same basic dental design and developmental process as their cousins; as vertebrates, they're made of bone and are bilaterally symmetrical and so on. They are, then, unique vis-à-vis dental morphology and killing style when described in terms of throat-clamps or reduced molars, but not in terms of placental molar development nor hunting techniques.

Comparative inferences are also *specific* to description. Making a phylogenetic inference, say, from a skull's being distinctively catlike to that critter being a member of *Carnivora,* tells us a limited range of things. The inference might allow us to claim the critter is quadrupedal, and likely that it is a predator, but more specific information about its killing style will not carry along the inferential pathway. This is even more pressing for convergent inferences. Identifying an environment-trait match in a lineage provides weak evidential support for me inferring from a lineage occupying a similar environment to it sharing that trait, but only specific to that trait (Currie 2013). Inferring from possessing a catlike jaw to being phylogenetically a cat, and thus possessing the set of typical placental and feline traits is defeasibly kosher, but inferring to hunting techniques is not. Specificity captures the defeasibility of comparative inferences. Particular details can reveal exceptions to comparative inferences and models.

The combination of sensitivity and specificity generates two epistemic problems. I'll call the first the *grain problem* (Currie 2012). If whether something is homologous or homoplastic turns in part on how the trait is described, then one can generate more or less homology or homoplasticity by shifting the grain of description. George McGhee's discussion of convergences, I think, sometimes falls prey to this (McGhee 2011). Consider his discussion of 'carrion-eaters':

> there exist some animals, the carrion-eaters, who have converged on the saprophytic, necrophagous mode of life. The corpse-seeking carrion beetles and hyenas are very different-looking types of animals, one an arthropod and other a mammal, yet they are ecological equivalents (McGhee, 144).

It is not obvious that these critters are 'ecological equivalents': their scavenging strategies differ and these differences could make a difference to how the communities they inhabit are disrupted by, or adaptively respond to, their presence. The concern then is that the category 'carrion eater' might not be ecologically meaningful. Further – as we'll see – McGhee's interest in convergence is in part driven by wanting to understand macroevolutionary patterns. He claims that convergence is ubiquitous, but this isn't so interesting if one can generate convergences by fiddling with grains of description.

The second issue we could call the *shallowness* problem (Griffiths 1994, Ereshefksy 2012). This is the idea that convergent similarities disappear at finer grains. As Paul Griffiths has put it:

> It is a truism in comparative biology that similarities due to analogy (shared selective function) are "shallow". The deeper you dig the more things diverge (Griffiths 2007, p. 216).

This is a pretty general claim, and there are sure to be exceptions. Regardless, the thought is that ancestry is typically a more powerful shaper of lineages than natural selection. While natural selection decides which forms survive and flourish, ancestry decides which forms are available for selection in the first place. Moreover, natural selection concerns functional categories, which can be realized in multiple ways; meaning that convergence often doesn't capture the fine details. Thus – so the thought goes – convergent inferences are of limited utility. Again, these problems become particularly pressing when we consider attempts to categorize and list convergences towards arguing for the repeatability of evolution at the macro-scale. These worries I think look much more problematic on the pluralist strategy, where we treat the evidence and explanations autonomously. I suspect the integrative picture sketched above relieves some of that pressure.

Together, the situated, sensitive, and specific nature of comparative methods can radically constrain what can be done with comparative data. Ancestry and adaptation jointly underwrite the power of such inferences, but both are processes that operate under specific circumstances, and integration – also necessary for legitimate handling of comparative evidence – brings these constraints together. It is in this sense that I claim comparative evidence is 'local'. Two points: first, I don't mean in saying this that other kinds of evidence are not local in some sense – perhaps senses rather similar to comparative evidence – but just that comparative evidence is *so* local. Second, by saying the evidence is 'local', I don't thereby commit to comparative evidence being unable to underwrite quite general models of evolution. The claim is rather that only under particular circumstances, when the situation, sensitivity, and specificity allow it, can they do so. Do they ever? That is, how general can comparative biology get? I suspect not very, and will argue for this in Section 3 when we tackle the shape of life. Further, are there circumstances where comparisons across biology can break free of locality? Specifically, are there laws of physics or chemistry which might provision claims across all biology regardless of phylogeny and the like? I suspect in some circumstances there might be (Russell Powell has recently provided an ambitious argument that there are; see his 2020), but my focus here will be on comparative thinking, not whether we can expand beyond it.

So, comparative evidence – and comparative thinking – has its limits. Nonetheless, from our discussion of the inferences involved, we can articulate the notion of a *paradigm comparative system*, that is, a system that is ideally suited to comparative thinking. As we saw, the stability of a phylogenetic inference turned on the fidelity of the phylogenetic signal. Roughly, how well preserved is phylogenetic information? This turns on both how strong the ancestral connections are – that is, the extent to which parental traits determine offspring traits – and whether that information is recoverable. The stability of a convergent inference turns on the tight connection between environment/trait matches – that is, to what extent natural selection explains the target traits and, again, how recoverable that information is. In both cases the recoverability of phylogenetic and convergent information will turn on both our ability to access that information and the number of examples. In some cases, phylogenetic information might be missing due to the lineage at hand being phylogenetically isolated. Unique or near-unique traits will lack analogue comparisons so even if the environment is extremely determining of the trait, the adaptive model will lack evidence, at least from that source.

Comparative thinking involves the integration of convergent and phylogenetic evidence, and so paradigm comparative systems will be rich in both. As we'll see later however, not being a paradigm comparative system is no block to comparative thinking being productive.

1.5 Styles of thinking

Ernst Mayr's characterization of what mattered about Darwinian conceptions of biology involved contrasting two styles of 'thinking' (Mayr 1959). One was *essentialist* or *typological* thinking, the other *population* thinking. At base, typological thinking involved conceiving of the living world in terms of fixed kinds, while population thinking conceived of the living world in terms of statistical variation. Mayr emphasized this shift throughout his life, still drawing on it at the turn of the millennium. He said of typological thinking:

> [Life's] seeming variety, it was said, consisted of a limited number of natural kinds (essences or types), each one forming a class. The members of each class were thought to be identical, constant, and sharply separated from the members of other essences. Variation, in contrast, is nonessential and accidental (Mayr 2000, 81).

And of population thinking:

> All groupings of living organisms, including humanity, are populations that consist of uniquely different individuals. No two of the six billion humans are the same. Populations vary not by their essences but only by mean statistical differences. By rejecting the constancy of populations, Darwin helped to introduce history into scientific thinking and to promote a distinctly new approach to explanatory interpretation in science (Mayr 2000, 82).

Although Mayr typically emphasized the metaphysical underpinning of population thinking, note his explicit appeal to mean statistical differences. The shift from typological to population thinking brings with it a commitment to a set of methodologies. Namely, tracing changes in mean variation in populations and modelling these using evolutionary machinery. In this sense, Mayr's conception of population thinking lies at the heart of the modern synthesis. The marriage of Darwinian evolution and Mendelian genetics provides the mathematical and theoretical grounding to track trait variation over time in heritability experiments, and to model those variations: that is, the study of microevolution. Of course, this is far too neat a history to be plausible. Joeri Witteveen (2016, 2018) has argued that Mayr's metaphysical gloss obscures a subtler methodological distinction in George Gaylord Simpson's work which leaves open space for typological thinking and methods. As Witteveen points out, the distinction is often used as a kind of rhetorical hammer in the service of methodological orthodoxy. Let's look at another style of thinking.

Recently, Marc Ereshefsky has introduced which he calls *homology thinking* (Ereshefsky 2012, 2007[9]). Homology thinking is focused, unsurprisingly, on homologies: 'Population thinking cites the structure of a population to explain the properties of a population. Homology thinking cites a character's history to explain its properties' (2012, 382). Instead of tracking shifts in averages in populations as Mayr would have us do, homology thinking involves tracing the history of a homologue across time. The jaw morphology of *Carnivora* is derived from the common ancestor of cats, dogs, and the rest. In one lineage – cats – a gap evolved between the molar and the incisor. Note that this involves the introduction of a type: the '*Carnivora* jaw'. But this type is not unchanging in the way that Mayr complained (Brigandt 2007, 2017). This kind of thinking has a standard kind of explanation: why do, say, dogs have the jaws they do? Or why do we see the similarities we do between the jaws of the various carnivores? That is because they're inherited from a common ancestor, and there is a set of methods and techniques applicable to this approach. Namely, mechanistic understanding of evolvability based on experimental studies of developmental systems, and studies of phylogenetics across time; a toolkit typically associated with evolutionary developmental biology.

What is a 'style of thinking' then? A style of thinking involves a paradigm kind of explanation, a set of related tools, and an associated, more ephemeral, 'perspective'. In population thinking, we explain the prevalence of traits in a population (and changes in prevalence) in light of those traits' fitnesses; our tools are Mendelian genetics, breeding experiments, and microevolutionary models; we conceive of life in terms of varied populations. In homology thinking we explain the presence of a trait in terms of its being inherited, and differences between lineages in light of their differing evolvability; our tools are experimental understanding of developmental systems and phylogenetics; we conceive of the living world in terms of shifting, varied, inherited kinds, or, depending on your metaphysical leanings, individuals. What about comparative thinking then?

Comparative thinking involves conceiving of the living world in terms of lineages which are shaped by ancestry and adaptation. The paradigm explanation integrates the role of natural selection and homologies to generate a narrative concerning a lineage or set of lineages: cats have the dental morphology they do, due to having inherited the *Carnivora* jaw and it being shaped towards efficient strangulation. The explanation, then, cites both developmental and selective factors. Such explanations often involve an evolutionary profile: the lineage in question is compared and contrasted with relevant others in terms

---

[9] Wagner 2016 uses the term somewhat differently.

of both the ancestry and the trait, as we saw with mammalian skull shapes. As befitting comparative thinking's integrative nature, the techniques and methods are varied, including optimality modelling, principal component analysis, phylogenetics, and so forth.

The central distinguishing feature between homology and comparative thinking is integration. Homology thinking is another example of the pluralist's strategy. Ereshefsky (2012) explicitly contrasts homology-based explanations with 'analogy' explanations, which appeal to the design features of traits. One might explain cat jaws in terms of the history of their ancestry, or one might explain them in terms of their suitability for strangulation. But comparative thinking explicitly integrates these perspectives via an evolutionary profile. No doubt, homology thinking overlaps significantly with comparative thinking: I'd be tempted to think of it as a subset or flavour of comparative thinking, but won't further consider their relationship here.

1.6 Biological Significance

I've described comparative evidence as explicitly limited: to the extent that the system of interest lacks a strong phylogenetic signal, and to the extent that it is not shaped by selection, comparative evidence will be weak. But comparative thinking matters for more than evidence, it also often matters for determining the significance of biological discoveries.

Some scientific discoveries seem more significant than others: knowing the relationship between atmospheric carbon dioxide levels and global average temperature is – perhaps – more significant than knowing the number of sand granules in the Sahara Desert. Ideas about what is scientifically significant matters for how biologists produce knowledge; when, for instance, is a discovery worthy of being published in a high-profile journal (or at all)? We might approach significance descriptively, asking how as a matter of fact scientists treat various discoveries or evidence. The answer here will be extremely complex: no doubt political, economic, and social factors – even fashion – play a large role. But we might also ask the question normatively: what *ought* we count as biologically significant? On this approach, we can distinguish between *internal* and *external* answers. The latter concerns what is significant about a discovery given what society at large (or some subset) want out of science (I take Kitcher's *Science, Truth and Democracy,* 2003, to be an example of an external approach). Understanding atmospheric carbon dioxide is significant because it is critical for tackling climate change. The former concerns what is significant about a discovery in terms of surrounding scientific knowledge: how does it bolster, challenge, or extend scientific understanding?

I think comparative thinking has something to tell us about internal answers to biological significance.

To see the role of comparative thinking in biological significance, I'll turn to a recent discovery: *Heracles inexpectatus* (Worthy et al. 2019). This is a recent find from St Bathans in the Otago region of Aotearoa (New Zealand). There's a site there producing a lot of fossils from the upper Miocene (16–19 million years ago) which have been transforming our conception of the history of Aotearoa's birdlife. *H. inexpectatus* is noteworthy, basically, due to its size: it stands a meter tall, easily double the previous contender for parrot-size-champion, their cousin the Kakapo (see Figure 9).

On an external conception of scientific significance I suppose *H. inexpectatus'* discovery might matter for being an enormous, charismatic animal. But comparative thinking provides an internal answer: the discovery reshapes and extends our understanding of various evolutionary profiles.

Aotearoa currently hosts three species of parrot: the arboreal Kaka, alpine Kea and flightless, terrestrial Kakapo (Livezey 1992). One question surrounding antipodean parrots is whether Aotearoa's parrot collection was seeded recently from Australia's much richer parrot fauna or whether they've a deeper, autonomous, history. Aotearoa was part of a much larger continent ('Zealandia') which broke away from Gondwana around 85 ma, before becoming mostly submerged. Over 25 million years, but particularly in the last 5, Aotearoa as we now know it was formed, mostly through volcanic activity. Did Kea and company arrive after

Figure 9 size comparison between H. inexpectatus and H. sapien (detail of fig. 1, Worthy et al. 2019) © Royal Society

the majority of Aotearoa rose from the sea or did their ancestors inhabit Zealandia? Phylogenetic work on Aotearoa's existing parrots suggest they are the remnants of a radiation in the last million years, but the St Bathas finds speak to a longer presence (Worthy et al. 2011; Worthy et al. 2017). A related question asks how diverse parrots in Aotearoa were. Fleming (1979) imagined a proto-Kaka which' radiated into the contemporary three lineages. Again, the St Bathans finds suggest a richer history reaching deep into the Miocene. In detailing the St Bathans biota, Worthy et al. (2017) link this history to Aotearoa's plants and climate. During the Miocene the islands were warmer, sporting tropical flora: 'far more complex than any shrubland-forest flora from the recent period in New Zealand [this] is undoubtedly what allowed a variety of parrots and pigeons to exist' (1114).

Another significant feature of *H. inexpectatus* concerns patterns of island gigantism. Island gigantism – the tendency for lineages inhabiting islands to adopt much larger forms than their mainland cousins – is common in birds, often involving flightlessness like in the Kakapo (Livezey 1992). *H. inexpectatus* is the first truly giant parrot. This tells us something about the potential morphology and evolutionary patterns within those birds. And this is tied into differences between patterns of island gigantism on larger and smaller islands:

> *Heracles inexpectatus* adds to the suite of insular birds that have evolved giant and often flightless forms. This phenomenon is not restricted in taxonomic scope, but instead occurs across a surprising spectrum of groups including palaeognaths, tytonids and accipitrids (Worthy et al. 2019, 4).

This allows Worthy et al. to add to our understanding of the evolutionary profile of *Aves* on islands. They point out that in smaller islands (volcanic ones, for instance) there tends to be a single lineage which 'spawned giant forms' (ibid, 4), however there is no pattern regarding which bird will do so. Roughly, they chalk this up to founder effects: the lineage that happens to turn up, radiates into that niche. By contrast, larger islands such as Aotearoa and Madagascar have a more varied set of niches, allowing for multiple birds to 'go giant':

> The NZ mainland is larger and more ecologically complex than most islands and, lacking mammalian predators, predictably has produced the greatest diversity of giant avians anywhere (ibid, 4).

Work on *H. inexpectatus,* and on the St Bathans fossils generally, have all the hallmarks of comparative thinking. First, the lineages are understood via integrating ancestry and adaptation. Driving questions include whether Aotearoa's parrots draw ancestry from their Australian cousins and include whether Aotearoa's shifting environments played a role. Second, finds are significant due to feeding into various evolutionary profiles. These include both profiles of Aotearoa's parrots in particular, and island gigantism across birds more generally. If Worthy et al. are right, then on smaller islands we should expect a single giant lineage, but who goes giant is unpredictable, turning on who happens to arrive on the island; on larger islands, with more diverse niches, we should expect multiple gigantic bird lineages. Third, the studies exhibit locality: inferences and explanations are explicitly constrained to within phylogenetic groups and turn on the details of the critters at hand. Finally, evolutionary profiles are parataxic – although the models of island gigantism we've discussed are restricted to birds, they do not apply across all birds, as some are more amenable to gigantism than others.

H. inexpectatus' case tell us something about biological significance: the giant parrot is significant because of how it transforms our understanding of evolutionary profiles. It shows us that parrots are capable of being true island giants, and that Aotearoa's past parrot diversity is significantly richer than we previously thought. Comparative thinking, then, gives us a local understanding of when and how a biological finding might be significant: due to how the particular lineage is situated in ancestry and adaptation. This is an instance of internal rather than external significance: the integrative, situational aspects of comparative thinking make *H. inexpectatus'* discover significant due to transforming surrounding biological knowledge.

Thus far, we've characterized comparative thinking: the integration of adaptation and ancestry in conceiving of lineages in terms of evolutionary profiles. In the next two sections, I'll apply these lessons across two contexts. First, in comparisons of cognition across taxa, second, in terms of macroevolutionary outcomes.

2 Comparative Cognition

My aim in this part is to show how comparative thinking makes sense of experimental practices aimed at the cognitive capacities of non-human animals. I'll show how the locality of comparative thinking, and the aim of generating evolutionary profiles, feeds into the research directions and epistemology of comparative psychology. I'll do this via an interpretation of 'Morgan's Cannon'.

H. sapiens' distinctive lifeways are owed at least in part to our capacity for problem solving, abstract thought, empathy and cooperation: our cognitive prowess. This motivates several research programs deeply embedded in comparative thinking: examining cognition across taxa, often with an eye to understanding intelligence in humans. There are plenty of examples of evolutionary profiles built around comparative cognition. For instance, Robin Dunbar and collaborators track relative brain size against social group size in primates

(Dunbar 2009). It turns out that brain size (relative to body size) positively corresponds to group size. Showing a statistical relationship between these perimeters motivates hypotheses about our own intelligence (the 'social intelligence' hypothesis), sets further comparative questions (does the correlation hold across other taxa?) and generally drives research (although there's reason to worry that brain size is too coarse a measure, Logan et al. 2018).

One approach to comparative cognition is explicitly experimental. Here, a cognitive trait is operationalized – made experimentally tractable – and examined across lineages. Some methods are better for cross-taxa comparisons than others. For instance, a common measure for 'creativity' in adult humans is Guilford's Alternative Uses Test (Guilford 1967). Here, a simple object (say, a brick or a paperclip) is presented to the subject and, within a set timeframe, they name as many uses for the object as possible. Coming up with alternative uses for an artefact beyond its standard function requires transcending 'functional fixedness', seeing an object *as that kind of tool*. Having a hammer and seeing everything as nails is one thing, but having a hammer and using it as a makeshift anvil goes beyond what hammers are typically 'for'. So, the number of alternative uses someone can name for an ordinary object serves as a proxy for their creative capacities. This is next to useless for subjects with no, or very limited, linguistic capacities. Here, new experimental studies need to be conjured, and we'll see some later.

This experimental approach has some tricky conceptual and epistemic issues. When different taxa do succeed in their experimental tasks, this leaves open what cognitive mechanism was involved in that success. Bluntly, in solving the task did they use the same fancy cognitive machinery that you or I would use, or did the subject 'cheat' somehow? The go-to example of 'cheating' is the horse Clever Hans, who was apparently able to provide the correct answers to simple arithmetic problems via hoof stomping. It turned out that as opposed to doing math, Hans was responding to subtle queues from the audience (everyone tenses up, imagine, when Hans comes to the right number of clomps, and he's learned to stop after that signal). To my mind, these both involve fancy cognition: I might well be better at adding and subtracting than I am at picking up social queues of that subtlety. Regardless, comparative psychologists use this as a paradigm example of a certain kind of anthropomorphic mistake: projecting humanly cognitive capacities onto non-human animals. To guard against this they appeal to a principle usually called *Morgan's canon*: in short, don't explain a behaviour with fancy cognition if hum-drum cognition will do the job.

The perspective provided by comparative thinking, I'll argue, brings insights to comparative cognition. Regarding Morgan's canon, the principle properly understood plays an important role in shaping experimental research in comparative psychology.

2.1 Morgan's Canon

'Lloyd-Morgan's canon' (or just 'Morgan's canon') was a late nineteenth-century principle which played a crucial role in shaping the development of psychological behaviourism in the first half of the twentieth century, and – depending on who you talk to – still plays a role in comparative psychology today. At base, the canon is a rule about what kinds of cognitive hypotheses should be preferred in explaining animal behaviour. The canon is often framed as avoiding anthropocentrism: interpreting animal behaviour via human-style cognition: theory of mind, intentionality, and so forth. Here's Morgan's formulation from 1903:

> In no case is an animal activity to be interpreted in terms of higher psychological processes if it can be fairly interpreted in terms of processes which stand lower in the scale of psychological evolution and development (Morgan 1903, 59).

The general thought being that if we prefer 'lower' psychological processes, then we'll be less likely to over-project from our human subjects to our animal friends (of course, over worrying about anthropomorphism can introduce foibles of its own, see Buckner 2013). Morgan's principle immediately raises a set of questions: first, what makes a psychological process 'higher' or 'lower'? Second, what does it take for something to be 'fairly interpreted'? And third, why would we commit to such a principle? On my view, Morgan's appeal to psychological evolution and development is key: we can read the principle as an instance of comparative thinking. In recent discussion, focus has been on two interrelated questions: what is the nature of the canon (is it, for instance, a kind of parsimony) and what justifies it. Comparative thinking, I'll argue, sheds light on both of these questions.

Broadly speaking, accounts of Morgan's canon either consider it a kind of parsimony or as a more empirical principle often mistaken for parsimony. 'Parsimony' is an epistemic principle which says, in effect, *more simple = more likely to be true*, which can be cashed out in many different ways (Sober 2009). On this approach, Clever Hans' reading of body language is a simpler explanation than he being able to perform arithmetic, and due to this simplicity the former hypothesis should be preferred. On a more empirical reading, the canon is akin to a summary of the kinds of expectations we should have about cognition. Horses being sensitive to human body language is less surprising than horses being able to count and do sums. The former reading is more amenable to *a-priori* treatments, and the latter of *a-posteriori* treatments,[10] but neither necessitates one or the other kind of justification.

---

[10] Given the philosophical whirlpool surrounding these terms, I hesitate to define them, but basically an *a priori* justification is in some sense independent of experience, whilst an *a posteriori* justification depends on experience.

I think we should dismiss *a-priori* treatments of Morgan's canon. As we've seen, comparative thinking in biology is an epistemically local affair: the epistemic powers of a comparison turn on how it is ancestrally and environmentally embedded. As such, our preferences for hypotheses are unlikely to take the general, abstract form favoured by *a priori* defences of simplicity. The living world is obstinately varied and resistant to general treatments. What we want, then, is an *a-posteriori* defence: can we ground expectations for simplicity in our knowledge of cognitive systems? Answering this question requires working out what we mean by 'simplicity'. There are two general approaches: a synchronic notion where simplicity is considered in terms of cognitive function; a diachronic notion which cashes out simplicity in terms of ancestry.[11] I'll defend a version of the latter.

Many treatments of Morgan's canon emphasize a *functional* conception of 'simplicity': 'a higher cognitive process is one that endows the animal with more elaborate cognitive capacities than does a lower process' (Fitzpatrick 2008, 226). We can understand 'elaborate' in terms of, well, doing more stuff. Simon Fitzpatrick uses first- and second-order intentionality to illustrate what he has in mind. Here's an example of first-order intentionality: 'I believe Morgan's canon is justified'; and here's an example of second-order intentionality: 'I believe that they believe Morgan's canon is justified'. For Fitzpatrick, second-order intentionality counts as a higher – less simple – cognitive process because it allows more stuff to be done:

> second order intentionality is a 'higher' cognitive process than first-order intentionality, since possession of second-order intentionality allows an organism to form mental states whose contents concern other mental states, while possession of first-order intentionality does not. (ibid, 227).

This definition of 'simplicity' differs significantly from 'theoretical' or 'ontological' parsimony (Nolan 1997). In that context, we're concerned with, say, the number of things posited in a theory, not the relative sophistication of cognitive processes. This makes it tricky to claim that Morgan's canon is simply the application of general simplicity considerations to the case of comparative cognition. Worse, and as we'll discuss further, sometimes in comparative cognition there are multiple notions of simplicity at play simultaneously. If we take these as applications of general rules, then it seems we have an empirical impasse.

Some philosophers have developed more local, empirical notions of Morgan's canon. Here is Dacey (2016)'s:

---

[11] 'Synchronic' is at a time-slice, 'diachonic' across time.

Morgan's Canon (Revised): when two or more models match the data, prefer the model that posits the simplest psychological process, if the complex process is not already thought to be present in the species, and there aren't overriding contradictory parsimony claims.

Here, the canon is overridable both by us having good reason to think the cognition is already available in the subjects (which is why positing second-order intentionality in *H. sapiens* is kosher), and when other kinds of simplicity are not on the table. Its applicability, then, is sensitive to local conditions. However, Dacey's notion of simplicity still tracks function, and I think this has real problems: let's turn to these.

Functional definitions of cognitive simplicity only make sense in a limited number of cases, certainly fewer than Morgan's canon is supposed to apply to. Consider Clever Hans. There, we compared the cognitive function of sensitivity to cross-species body language with the capacity to do arithmetic. In what sense might the latter involve 'doing more stuff' than the former? Certainly, I intuitively 'see' that mathematical cognition is more impressive than reading body language, but I'm not sure how to capture this intuition and indeed worry that it is explained by general biases towards abstract thinking as a mark of intelligence. Further, I'm not sure how to count cognitive functions: one capacity enables you to do things the other does not and *vice-versa*. In sum, I think functional accounts look plausible when the cognitive capacities compared relate in a systematic way, such has having a belief versus having a belief about a belief, but they become much, much less sensible when we are comparing very different kinds of cognitive capacities, such as performing simple arithmetic and reading body language. Further, as Elliot Sober (2005) has pointed out, evolutionary conceptions of life are not friendly to the 'higher' and 'lower' talk functionalism implies:

> The theory of evolution by natural selection undermines the idea of a linear scale of nature in which each stage is higher or lower than every other. Darwin replaced the ladder with the tree; lineages diverge from each other and develop different adaptations that suit them to their peculiar conditions of life. In this framework, it makes no sense to ask whether one contemporary species is higher or lower than another (ibid, 91).

The idea that functionally less sophisticated cognition should be expected over and above more sophisticated cognition (even in a defeasible, local way) seems to assume a kind of directionality to evolution: that it shifts from simple, low function to complex, high function. One way we could defend this directionality is to point out that according to many experiments simple associative cognition is taxonomically widespread. But as Irina Meketa (2014) has pointed out, these experiments were themselves performed guided by Morgan's canon so are potentially biased towards simple cognition in the first place. Further, the arrow from simplicity to complexity cannot be taken for granted: there are examples working in the other direction too (Sober 2006; O'Malley 2014). The natural response to these qualms is to say that impressive cognition must be built from less impressive cognition, that is, some kinds of (simpler) cognition might be pre-conditions for other (more sophisticated) cognition (see Karin-D'Arcy 2005). I think there is something right there, but this masks a kind of anthropocentrism: that the *most* sophisticated cognition is human cognition, and so the preconditions of human cognition are, well, the cognition other animals possess. But why think this? It is possible – I think likely – that we are *less* good at reading some kinds of body language than Clever Hans was. That is, there may be forms of cognition that are more functionally sophisticated than our cognition in some ways, while ours is more functionality sophisticated in other ways. In that scenario, although different forms of cognition might have preconditions, there might be no general set of preconditions, nor general set of sophisticated cognitive functions.

In short, functional analysis of cognition falls afoul of the locality of comparative thinking. In the next section I'll develop a different approach.

2.2 The Comparative Canon

Accounts of Morgan's canon focused on cognitive function often start with an earlier summation of the canon:

> In no case may we interpret an action as the outcome of the exersize of a higher psychical faculty, if it can be interpreted as the outcome of the exercise of one which stands lower in the psychological scale (Morgan 1894, 53).

Missing Morgan's 1903 reference to 'evolution and development' (see earlier) has, I suspect, led these accounts to lack a comparative perspective.[12] I'll argue that we should take Morgan's appeal to psychological evolution and development seriously. To do so, I want to explore a recent example of Morgan's canon-style thinking and develop an account from that.

Tides are challenging for fish that live close to the shore: if you're caught out by the low tide, you're in trouble. One solution is to adopt a dual-breathing strategy, getting oxygen from water and air. Another strategy is to flop oneself to the safety of a tidepool, salt-water havens left in the wake of the withdrawing tide, or back to the sea itself. Mudskippers, mummichogs (or killifish), and Frillfin Gobys all adopt some combination of these strategies.

---

[12] One exception is Sober (2006) who directly ties Morgan's Canon to phylogenetic concerns.

Frillfin Goby have been the focus of some of the most famous experiments in fish cognition. Threatened gobies are able to leap from tidepool to tidepool. How do they know where the tidepools are? From the 1950s to 70s Lestor Aronson tackled this experimentally. Divide a bunch of gobies into an experiment and a control group. The experimental group is allowed to swim about an artificial sea at high tide, potentially gaining information about the contours of the seabed, while the control group is not. The performance of stranded fish is then compared at low tide: are the experimental group more accurate jumpers? Apparently so, and this suggests a particular explanation for goby success: they are literally exploring, mapping, and then remembering the sea floor. But how do they manage this task? As Brown, Laland, and Krause have put it:

> What types of landmark do the fishes use to recognise which particular pool they are in – local or global? Do the fishes remember a sequence or jumps, or are they able to make an appropriate escape response based on where they are at that point in time? (Brown, Laland and Krause 2008, 169)

So, we're asking how gobies orient themselves. Compare two hypotheses. On one account, fish use 'global' cues, such as the dappled reflections of light on water. However, in Aronson's experiments, the gobies would only leap between familiar pools, and their leaping behaviour couldn't seem to be explained by detection as, first, they couldn't see into the next pool, and second, they would leap into empty but familiar pools. By another hypothesis the queues are local, that is, the fish learn particular landmarks about their local environments:

> These data can be interpreted as indicating a remarkable instance of latent landmark learning: The jumpers may have learned the terrain and arrangement of tide pools in their home area, presumably by swimming over the area during high tide (Hoar and Randall 1976, 203).

But as Hoar and Randall point out:

> further replication and investigation of this phenomenon may reveal a simpler mechanism (203).

Although Hoar and Randall reckon the experiments favour local queues, they also clearly think we should prefer global queues – 'simpler mechanisms' – if we can.

Reflecting on global and local hypotheses of fish orientation give us some handle on Morgan's canon. Perhaps 'global' hypotheses involve Morgan's 'lower' cognition, while local hypotheses implicate 'higher' cognition. But, given that I don't like functional readings, how should we interpret 'higher' and 'lower' here? I think the most plausible reading of 'lower' is phylogenetic: global hypotheses involves traits which are 'primitive' in the neutral sense of emerging earlier in the tree of life – and thus being present across large groups of taxa. In addition to this phylogenetic aspect, I also think a notion of *evolvability* is implied. Rules such as 'move towards things that reflect light' are plausibly much easier to evolve by co-opting already existing machinery to new tasks (such as light detection and behavioural adjustments sensitive to those queues). By comparison, local hypotheses require much more information gathering, storage, and then retrieval. Mapping out a seabed and then remembering it requires more – as it were – memory and RAM than performing a single action no-matter-what (if you'll forgive the loaded computational metaphor). But it isn't that it requires more RAM *per se* that makes it unlikely, but because there is good reason to think that such cognitive machinery is evolutionarily expensive: brain tissue consumes a lot of resources, so if there are less metabolically demanding ways of solving the problem selection will favour them. A final point about evolvability involves the selection pressure mummichogs and guppies have likely undergone in the past. Given the extremely high cost of being stranded, it's a plausible conjecture that selection pressure for mitigation was high. We've seen over and again how natural selection under the right circumstances can produce apparently very sophisticated behaviours from simple, but highly adaptive, rules. When considering a behaviour under so much selection pressure we should expect ingenious solutions – and these need not involve human-like cognition.

I've suggested we read Hoar and Randall's handling of global and local fish orientation in a comparative context. That is, the global hypothesis should be preferred because it explains the behaviour by appeal to a capacity which is found across many taxa (and so is phylogenetically 'deep'), which is relatively inexpensive metabolically, and is a solution for an important problem in an evolutionary context. In humans we already know that expensive, fancy cognitive machinery is available, so these considerations do not arise. In the fish's scenario, we have bountiful reason to expect responses to global queues, but not local queues. This is classic comparative thinking: the target is situated in an integrated evolutionary profile. We can generalize this analysis:[13]

> We should prefer cognitive explanations which (1) appeal to cognitive capacities we already know are present in the critter or (2) appeal to cognitive capacities which are highly evolvable insofar as (a) they are easily cooptable from cognitive capacities we already know are present in the critter or (b) for which there is high selection pressure.

---

[13] Fitzpatrick mentions some of these comparative notions (2008, 233) however uses these as reason to reject Morgan's Canon, as opposed to the basis for a differing account. Dacey (2016) also summarizes defences of Morgan's Canon which appeal to evolutionary considerations.

Note the conditions (and subconditions) are disjunctive, meeting them all is unnecessary. I don't think there is a general answer to be had about when and in what combination they are required: like so much in comparative biology this will depend on local context. The justification of the principle lies in Section 1's account of comparative thinking. To see this, let's apply the analysis to the Clever Hans case. There, the two hypotheses were *Hans can count* and *Hans can read subtle queues in human body language*. The first condition can be met through observation or by phylogenetic inference. Focusing on phylogenies, sophisticated social interaction and bodily communication is a common feature of mammals (consider the almost ritualistic behaviour of house cats resolving territorial disputes). Via a phylogenetic inference, then, we should not be surprised by cross-species bodily sensitivity in mammals, particularly social ones like horses. Regarding the second condition, although it is unlikely that all mammals have evolved to be attuned to human behaviour, it isn't so remarkable to think it fairly evolvable. After all, there are examples of cross-species sensitivity to behaviour in other mammals (consider predator avoidance, or the apparent cooperation between herd animals on the Savannah). More critically, as domesticated animals, horses have been under extremely high selection pressure to be sensitive to human behaviour. By comparison, evidence of mathematical acumen in mammals is fairly scant (birds are stand-outs in that regard, Emmerton 2001). Additionally, there is no obvious route from common features of mammalian cognition to arithmetic capacities, and as far as I'm aware there's not much selection pressure on horses to do sums.

In my view, then, Morgan's canon is best understood as shorthand for applying comparative thinking to animal cognition. As such, it is no version of parsimony – we're not appealing to 'simplicity' here (and nor, you might note, did Morgan in his summations). Further, it is justified empirically, not on *a priori* grounds. Indeed, its justification is highly local in the sense I discussed earlier. It is highly sensitive to, and revisable in light of, comparative context. That we prefer global rather than local problem solving in fish is due to our background knowledge about ancestral traits in fish and the selection pressure on them. Shifts in that knowledge could shift the balance one way or the other.

Let's defend the account against common objections to Morgan's canon, before considering its role in experimental practice.

First, you might think this just isn't Morgan's canon. You might complain that, as I'm not talking about a general methodological principle (or at least the right kind of principle), I've illegitimately changed the subject. There are at least two versions of this challenge. One is historical: have I isolated what Morgan had in mind when he put forwards his canon? I doubt that I have, although Sober (2006) pays attention to Morgan's historical context and also prefers a version grounded in comparative thinking. However, I'm not focused on the otherwise interesting question of historical interpretation (see e.g., Radick 2000). Another challenge asks whether my analysis is what scientists have in mind when they state allegiance to (or against) the canon. Although this is closer to my approach, I am not engaging in a kind of conceptual linguistics of comparative psychology. My aim is not to describe and systematize scientific usage. Let's call my approach *conceptual pragmatism*: one examines the practices at hand, and asks what concepts will best do the work those practices demand (similar in spirit to Mitchell 1997; Haslanger 2000). This is beholden to how scientists talk, but only partially. Its success turns on whether I've laid bare the patterns of justification which underlie scientific practice – does it explain why scientists do what they do and, hopefully at least, provide insights regarding whether they ought to? Good analyses on this approach might well depart from what scientists explicitly say, think, or write.

Having said this, I do suspect the comparative canon often *is* what comparative psychologists have in mind when they refer to simplicity. Scientific talk often hides highly sophisticated, often tacit, concepts and practices behind apparently monistic terms (on gene concepts, see Waters 2007). As such – and this is no more than a bold hypothesis – by 'simplicity' they might well mean the comparative canon. And if they don't, I've argued, they should. There might still be some reasons to give up on the term 'Morgan's canon'. Perhaps the term is too associated with the functional, hierarchical accounts rejected in the last section; or there could be historical baggage: the canon could be associated with wrong-headed science that has been, or should be, abandoned. With these caveats, however, I'm happy to continue usage as I have.

Second, others object to the canon due to clashes between virtues (Dwyer & Burgess 2011; Dacey 2016). Does 'simplicity' track the number of cognitive processes, or does it track the cognitive complexity of processes? As we saw above, such clashes led Dacey to add a rider to their account of the canon, telling us to prefer simplicity unless other factors were at play. On my account these apparent clashes boil down to countervailing evidence and background knowledge. Thinking of the canon in terms of parsimony makes for an apparently unresolvable clash: which kind of simplicity should I prefer? But on my account such debates are tractable and can be understood as differing evidence for the nature of the ancestral and adaptive context. This doesn't mean they will always be resolved (that turns on empirical fortune) but such challenges arise in empirical investigation all the time.

Third, Irina Maketa convincingly argues against grounding Morgan's canon in appeals to taxonomy: trying to defend a preference for simplicity in life's ancestry falls afoul of the diversity of life's paths (Maketa 2014). I think Maketa is right, but these considerations do not hit my partly-taxonomic conception of Morgan's canon. Maketa is objecting to arguments from phylogenetics to cognitive simplicity. I am in a sense interpreting 'simplicity' as meaning (in part) 'phylogenetically primitive'. That is, I am not defining Morgan's canon functionally, and then making a phylogenetic inference to suggest we should expect functional simplicity. I am rather defining the canon in phylogenetic terms.

2.3 Pursuit in Comparative Psychology

The fourth objection I want to consider appeals not to whether Morgan's canon gives us good reason to believe one hypothesis over another, but whether it grants good reason for pursuit. That is, is following Morgan's canon a productive strategy?[14] Simon Fitzpatrick thinks not, claiming that it has 'pernicious implications for research in animal psychology' (2008, 236). Fitzpatrick doubts the pursuitworthiness of the canon because it leads to what he calls the 'Scaling Up Strategy':

> *Scaling Up Strategy*: In any given area of research into animal behaviour, the working hypothesis should always be the least cognitively sophisticated explanation consistent with the available data (ibid, 236).

Fitzpatrick argues that such a strategy works against the critical role of speculation in driving cognitive science. 'In any instances unless one is prepared to speculate far beyond the current data, to seriously entertain hypotheses involving more sophisticated cognitive processes when a less sophisticated one seems sufficient, it is unlikely that one will ever uncover such subtle and surprising capacities' (238). Elsewhere, I've stressed the importance of speculation in driving scientific advancement (Currie 2018, 2019c), so the objection is very pressing. Further, I've argued in Section 1 that comparative thinking plays an important role in shaping the direction of biological work: that developing an evolutionary profile often explains what biologists see as significant. If Fitzpatrick's objection holds against my version of Morgan's canon, we'd be tempted to say that such notions of significance are deleterious. Happily, I think the objection is ill-founded, in essence because it conflates two ways of understanding the canon. First, telling us which hypothesis to believe or not; second, telling us which kinds of confounders to consider in experimental design. Distinguishing these roles, I'll argue, allows us to see the canon's fruitfulness.

Where philosophers (and some scientists) interested in Morgan's canon have focused on *acceptance* – when should we think a hypothesis is true or not? –

---

[14] See Nyrup 2018, Currie 2019a on the context of pursuit in science.

some historians have emphasized the connection between the canon and experimental practice. Greg Radick, for instance, connects Edward Thorndike's development of puzzle boxes in the early twentieth century to Morgan's work. Puzzle boxes involve trapping an animal in an enclosed space, their escape requiring performance of a series of actions in a particular order. By trapping the same animal on consecutive trials in the same box, and measuring their escape times, Thorndike was able to generate a 'learning curve'. Roughly, the animals were learning by trial and error, and then remembering their solutions on future runs. According to Radick, in such experimental paradigms:

> Morgan's canon found belated instrumental embodiment. Using the puzzle boxes, Thorndike was able for the first time to quantify the exercise of 'intelligence' in Morgan's sense (Radick 2000, 21).

There is a lesson here. Instead of asking whether following Morgan's canon will grant good reason for belief, let's instead ask how it instructs us to conduct experiments, the bread and butter of comparative psychology. Later, I'll delve into experiments on mummichogs and their capacity to orient themselves towards water when stranded. One experimental intervention attempts to 'fool' mummichogs by using tinfoil as source of reflectance. The aim is to control for a confounding variable by testing between the hypothesis that mummichogs recognise water, or merely reflected light. Examining naturally jumping mummichogs will not differentiate between these, as in their natural environment water is the only source of such light. The point of the experiment is to empirically differentiate between the two hypotheses. Considering Morgan's canon, the tinfoil experiment makes perfect sense. Recognising patterns of light has a deep ancestry across animals – indeed, it is plausible that differentiating between light levels is the original function of the light-sensitive-spots from which eyes evolved. So, recognising that kind of phenomena is likely highly evolvable (and especially so given the selection pressure on fish like mummichogs to escape being stranded).

As we saw, Fitzpatrick's scaling-up strategy was about which hypothesis should be our 'working hypothesis' (perhaps similarly to a 'pseudo-null' hypothesis in Bausman & Halina's 2018 sense). Although his objection is geared towards pursuit, his principle isn't – it is about which hypothesis we should prefer (see also Fitzpatrick, forthcoming). Instead, we could seek Morgan's canon's justification in terms of how it directs experimental study (Zentall 2017).

Distinguish between two tasks: identifying confounding variables and testing a hypothesis. The latter asks whether we have good grounds for accepting or rejecting some claim about a system of interest. 'Mummichogs use global queues' for instance. The former asks whether a particular experimental design has isolated the right variables. 'This experiment distinguishes between Mummichog's capacity to land in water being due to water-recognition or recognising reflected light' is an example of the latter. The identification of confounding variables leads to tweaks and developments in experiments, such as trying to 'fool' the mummichogs with tinfoil. My identifying particular confounders for a particular experiment does not immediately decide my beliefs about the hypotheses the experiment is designed to bear on. I might, for instance, be very confident that mummichogs use global reasoning, but still recognise that vis-à-vis that experimental set-up, the experiment doesn't do the distinguishing work I want it to. In other words, when considering the accept-ance or otherwise of a hypothesis, we consider a wide range of evidential (and perhaps extra-evidential) factors. When considering confounders we ask whether this particular experiment can distinguish between two hypotheses relating to how subjects behave under experimental conditions. Now consider again Fitzpatrick's 'scaling-up-strategy'. On that account, Morgan's canon tells us that the 'working hypothesis', the hypothesis we will prefer to accept, is the simplest (copy in your favourite conception of 'simplest' here). Fitzpatrick argues that this puts a break on the speculative hypothesizing which is so important for scientific progress. But once we split acceptance from identifying confounders, we see a possible response to Fitzpatrick: if in experimental contexts we should understand Morgan's canon as telling us how to identify confounders, as opposed to which hypotheses should be our base-line expect-ations, then the scaling-up strategy doesn't follows from the canon. This is because, for instance, on comparative grounds I might expect highly sophisti-cated cognition in my subjects. When considering which hypothesis to believe, I think my analysis of the canon has some legs, and also for identifying confounders it has benefits. But these are separate activities, and Fitzpatrick's argument conflates them.

With this distinction, we now have two questions regarding Morgan's canon. First, does following the canon provide good reason for accepting or rejecting claims in comparative psychology? Second, is following the canon a good way of productively identifying and controlling for confounders? I think we can answer both questions affirmatively. On accepting claims about the cognitive prowess and mechanisms which non-human animals possess, Morgan's canon tells us that we should prefer hypotheses given local phylogenetic and other evolutionary information about the critter at hand. For a case like Clever Hans, mathematical cognition (and language recognition) is not something we expect in mammals like horses – there is neither strong selection for it, nor is it present in the phylogeny nor clearly evolvable from it – but close attention to body language and social clues is. This doesn't stop us from speculating, of course: but speculation is not the same as theory acceptance (Currie 2019b; Turner 2019). On this understanding, Morgan's canon is not a rule about what kind of hypothesis we should consider or experiment with, but a rule about what kind of hypothesis we should expect to be true.

Now, consider pursuit and generating confounders. Morgan's canon tells comparative psychologists what to look out for in their experimental design and how to proceed in experimental treatments. Let's look at a few more studies of fish cognition. As we've seen, various hypotheses might be produced for how oft-stranded fish make their way back to safety. Way back in 1915, S. O Mast noted the general behaviour we're interested in, reporting remarkable terrestrial excursions amongst minnows (Mast 1915):

> I have seen more than 200 of these fishes leave a tide-pool 50 meters long, 13 meters wide and 30 cm deep, and travel across a sand-bar more than 3 meters wide and 10 cm high, all in the course of half an hour. And I have seen them proceed in a fairly direct course toward the sea even against a moderately strong wind. I have also seen them attempt, continuously for at least a minute, to go overland to the sea against a wind so strong that they could make no headway. When I first saw this performance I was deeply impressed. I had often seen fishes, when thrown on the land, flop back into the water in a more or less aimless fashion, but I had never seen any voluntarily leave a body of water and travel in a coordinate way on land (ibid, 345).

Minnows will, on being trapped in tidepools as the ocean recedes, escape together towards the sea. Mast emphasized the minnow's tendency to escape via the outlet that feeds the tidepool as the tide goes out. As they make their escape after the outlet has closed, Mast supposes '[t]he location of this outlet they evidently remember for some time' (349). Some of Mast's experiments involved attempting to block the minnow's vision of the sea with pieces of card. As this didn't change their behaviour, he concluded that 'The phenomenon is consequently probably very largely dependent upon internal factors' (350). By 'internal factors' Mast meant that the minnows map their tidepool and remember where it is.

Using a different fish, Bressman, Farina, & Gibb (2016) explore a contrasting explanation. Fish have several ways of generating terrestrial momentum. Mummichogs perform 'tail-flip' jumps. Momentum is generated by 'rolling over on their lateral side, bending their heads towards their tails, then pushing off with their peduncle in a caudally-orientated jump' (60). The peduncle is the fleshy part of the tail before the fin. Typically landing on their side, they then position themselves in an 'upright' position, before flipping over and repeating the pattern:

> There is a cyclical nature to the terrestrial behaviour and locomotion of mummichogs: they perform a tail-flip jump land on their side, perform a reorientation behaviour to bring themselves to an upright position (sometimes they wriggle and reorient the body in a different direction), roll onto their side, and jump again, restarting the cycle (Bressman, Farina, & Gibb 2015, 61).

Focus on a particular aspect of behaviour: mummichogs regularly stopping in an upright position. A tempting hypothesis is that the position is ideal for looking for water.

Bressman, Farina, & Gibb's experiments involved placing mummichogs on a square dry table, on one edge of which was a body of water, while the other edges involved drops into hidden buckets of water. On being released, the mummichogs begin their distinctive tail-flip, followed-by-uprighting routine. If their jumping were random, then there would be equal chances of any particular fish falling off each side of the table – so 25 per cent for each side. However, approximately 50 per cent of those who made it to an edge picked the aquatic side. This matters for demonstrating that the mummichog's behaviour is non-random, but it also pushes against Mast's take on minnows. Because they are naïve to the test conditions, the fish cannot have mapped and remembered the location.

Conducting further experiments – including the tinfoil one mentioned above – led Bressman, Farina, & Gibb to conclude that mummichogs are locating water by recognising the kind of reflection given off by water (and tinfoil!), and that the uprighting behaviour 'allows mummichogs to visually survey their environment' (63). As they say,

> it was possible that [Mast's] fish had memorized their habitat during high tide … and used a mental map to navigate to the ocean. However, the mummichogs in this study were able to navigate from an unfamiliar terrestrial environment to an unfamiliar body of water, so it was impossible for them to use a mental map to navigate towards water (62).

Bressman, Farina, & Gibb's argument doesn't simply depend upon experiments: it is also situated within comparative knowledge of fish. For instance, most fish (with some highly-adapted exceptions) are unable to focus on objects in the air (their eyes are adapted to aquatic mediums) and so, although vision is a central characteristic of fish, it is of limited use in terrestrial environments. So, we should expect in the typical fishy repertoire that they'll 'rely on simple visual cues, such as reflections of light off the water' (63). They also note the existence of several fish, like mudskippers, which 'prone-jump', that is, instead of the mummichog's rather involved back-and-forth from lateral to vertical positions, they jump from the prone position. This suggests convergence: 'The combination of tail-flip jumps and the uprighting behaviour of mummichogs may be a less-derived, yet analogous behaviour to prone jumping' (63). They hint that an adaptive model based on convergent evidence might be available and close by linking this to a story about how such behaviour evolves:

> For mummichogs, mangrove rivulus, possibly many other species of small fish that live at the water's edge, the uprighting behaviour may allow the fish to gather critical visual information about its environment before it initiates its next jump. The presence of such behaviors among disparate groups of amphibious fishes and the evidence presented here documenting the import- ance of visual clues for *F. heteroclitus* support the evolutionary hypothesis that a prone body position evolves as fish move onto land because vision is a critical sensory system for navigation in terrestrial habitats (63).

If you've bought Section 1 on comparative thinking none of this should be a surprise. Biologists situate their subjects in the relevant evolutionary informa- tion, constructing an evolutionary profile. This involves comparative work which identifies typical capacities within that homologous group (fishes typic- ally cannot focus on terrestrial objects), consider adaptive divergences from this (those fish which have developed such capacities), link the subject with poten- tial analogues (prone-jumpers such as mudskippers), and use these to construct evolutionary narratives. But what we're interested in here is how comparative thinking, as encapsulated in my development of Morgan's canon, guides experimental design.

If we construe Morgan's canon as about identifying confounders, it looks something like this:

> When considering an experiment probing some animal behaviour, consider confounders which involve (1) appeal to cognitive capacities we already know are present in the critter or (2) cognitive capacities which are highly evolvable insofar as (a) they are easily cooptable from cognitive capacities we already know are present in the critter or (b) for which there is high selection pressure.

Although physiologically most fish are unable to form images of terrestrial objects, vision is already an important feature and under special circumstances fish do adapt air-based vision. So we've got good reason to think vision is fairly evolvable in fish, and a likely locus for adaptive solutions to problems – especially in cases where there is high selection pressure, as there surely is for poor stranded mummichogs. By Morgan's canon, this makes global visual cues good candidate confounders. Another idea, for instance, might be that there is some pattern in how tidepools drain which mummichogs have adapted to recognise and react to. We don't have much evidence of fish adapting to landmarks in quite that way, although it is of course possible! So, my proposal is that what explains Bressman, Farina, & Gibb's experiment is that it is driven by the recognition that visual-hypotheses in fish are a good kind of confounder for experiments testing mapping-hypotheses. This can be true even if the experiment gave a false result. If, for instance, the mummichogs had performed at random or worse vis-à-vis locating the water, this would provide strong negative evidence against visual hypotheses.

So the canon can be a source of confounders, but why think it a good source? One reason involves tying the belief-forming version of the canon to the pursuitworthiness version. If the canon provides reasons to expect some hypotheses to be more likely than others, then that is a pretty good guide, one would think, for pursuit. I happen to think that the two versions of the canon are linked in this way, but I don't think an argument in favour of pursuit-worthiness must make appeal to that connection. A complementary approach points out that those features which are widespread in the lineage at hand are likely to have already been studied, or that studying them are more likely to have broad upshots across the group studied. Because the feature at hand is wide spread, there is a better chance of developing experimental traditions involving further related subjects: we could, for instance, expand the mummichog experiments to minnows. Moreover, comparative thinking involves building an evolutionary profile of our target: understanding what is, for the lineage at hand, evolvable, adaptive, developmentally open, and so forth, and setting this in a comparative context with other lineages. Because Morgan's canon (on my reading) is explicitly comparative, considering confounders and designing experiments following it should build just the know-ledge-base required for evolutionary profiles. This is not, however, a complete reply to Fitzpatrick: no doubt the canon still seems to restrict speculation. I've two further comments on this. First, it is not obvious that unrestrained speculation is so productive in science: it is instead the tight connection between evidence, background knowledge, and empirical investi-gation that underwrite productive speculation (Currie 2018). Second, espe-cially in its pursuitworthiness form, the canon acts as a productive guide, not an overweening law.

Let's summarize. I've argued that the best understanding of Morgan's canon involves comparative thinking. The canon is not about simplicity vis-à-vis some abstract a-priori principle, nor about functional cognitive capacities *per se*. It is instead about what patterns of cognitive prowess and mechanisms we should expect in critters of that phylogenetic and ecological history. I've made some-thing of a bold bet: that uses of 'simplicity' amongst comparative psychologists track something like my account. Further, I've suggested that arguments against the canon sometimes conflate hypothesis acceptance with identifying and controlling for confounders. For the former question, the canon (correctly) asks us to situate our understanding of animal cognition in a comparative context. For the latter, the canon (productively) asks experimenters to seek out and control for confounders from cognitive capacities expected from the ancestral-adaptive context.

Does this count as a blanket vindication of Morgan's canon? No. As we've seen, comparative thinking is at its most powerful when ancestry and adaptation are strong forces in shaping our target of interest. And this might not be true for all targets. It is unclear, for instance, whether we should apply this kind of thinking in artificial contexts, although engineering and evolution are sometimes analogous (Calcott et al. 2015). The locality of comparative thinking, its demand that we understand lineages in their phylogenetic and ecological contexts, makes the transfer of inference rules between biology and engineering tricky. In other words, no doubt we should strive to build comparisons between artificial and biological intelligences, but I suspect this will require leaving comparative thinking – and its power and insight – behind.

I want to close by highlighting a more subtle anthropocentrism which the comparative canon guards us from. As we've seen, many discussions of Morgan's canon emphasize the need to control for over-projecting human cognitive powers to other animals. No doubt I would recognise water by recognising it as water, but this doesn't license thinking the same of a mummichog. Morgan's canon is sometimes thought of as introducing a new bias to counteract the anthropocentric one. On my interpretation, we do avoid this kind of over-attribution, but not because we've adopted a new bias. Instead, our empirically grounded expectations about the lineage at hand do the work.[15] But anthropocentrism smuggles in another kind of bias often reinforced by functionalist conceptions: that there is a single, systematic, hierarchical way of understanding cognition (with our capacities placed somewhere near the top). As I speculated regarding Clever Hans, I see no reason to understand cognition as a monolithic nested hierarchy of dependence; it could prove to be a significantly more disparate, disunified, and messy business: and if we're to understand it, comparative perspectives, with all their locality, are essential.

The account of comparative thinking I provided in Section 1, then, can be bought to bare in understanding the inferential and experimental practices of comparative psychology. In grappling with the cognition of animal minds, scientists often think comparatively: situating their subjects within evolutionary contexts and iteratively building evolutionary profiles.

---

[15] In a sense, this aligns with Kirkpatrick's 'evidentialism'. However, where Kirkpatrick's account appeals to general features of evidential reasoning, mine is closely tied to distinctively biological reasoning, that is, comparative thinking.

3 The Shape of Life

Having considered comparative thinking in the concrete, experimental domain of animal cognition, let's switch gears and see how it plays out on a grander scale: understanding macroevolutionary patterns. Here, we again see the importance of comparative thinking, but we also confront its limits.

We've considered the biological significance of *Heracles inexpectatus*. The discovery of a giant extinct parrot enriched our conception of the ancestry of Aotearoa's birds and helped understand patterns of gigantism across *Aves*. It thus informed several evolutionary profiles: one concerning *H. inexpectatus* and her relatives, another about how birds adapt to island life. Regarding the latter, Worthy et al. (2019) said:

> It seems likely that the stochastic although often-times directional nature of successful dispersal and competitive exclusion by original founder species is what constrained the evolution of giant birds on smaller isolated islands . . . The NZ mainland is larger and more ecological complex than most islands and, lacking mammalian predators, predictably has produced the greatest diversity of giant avians anywhere (Worthy et al. 2019, 4).

Worthy et al.'s claims can be understood using a metaphor of sorts: *replays*. A 'replay' involves winding back evolution's clock to sometime in the past and then running it forwards again, asking which events will reoccur and which will not.[16] If we reset the biological clock to prior the dispersal of birds and play it forwards, what should we expect to see? If Worthy et al. are right, we should expect smaller islands to be colonized by single-lineage giants, yet we won't be able to predict which lineages in particular will form founding populations. Meanwhile in Aotearoa we'll still see various radiations of giant flightless birds. These claims are about the *shape of life*. How contingent is life's path? Will we see the same kinds of results over and again, or will there be surprising outcomes? Asking after the shape of life is to ask after the robustness or fragility of macroevolutionary outcomes.[17] As Lindell Bromham puts it, 'the study of macroevolution focuses on changes in biodiversity over time, space and lineages, describing and explaining changes in the representation of lineages in the biota' (Bromham 2016, 48; see also Grantham 2007; Turner & Havstad 2019). The dispersal of parrots across Aotearoa involved shifts in biodiversity across space in time as they radiated into new niches and Aotearoa's environments changed around them. We're here interested in the modal properties of such

---

[16] Gould thought of replays in terms of rewinding evolution's tape, rather than its clock. Whether this shift in metaphor makes a difference I leave as an exercise for the reader.

[17] Here by 'robustness' I mean the outcome will occur across a wide range of possibilities given certain starting conditions, while fragility is the inverse.

macroevolutionary events: did Aotearoa's parrot dispersal have to happen that way, or could it have been different?

Motivated by Stephen Jay Gould's *Wonderful Life* (1990), the shape of life is often treated very generally. Consider the *Radical Contingency Thesis* (RCT). At a first pass, this says that on replays the biological world would turn out vastly different than it has; evolutionary outcomes are fragile. The contrasting thesis is radical convergence: on replays the same kinds of events will reoccur; evolutionary outcomes are robust (see Beatty 1996; Powell 2009; McConwell 2017; Wong 2019). Russell Powell and Carlos Mariscal characterize the contingency thesis as follows:

> the [radical contingency thesis] is fundamentally a metaphysical (modal) thesis ... it holds that certain macroevolutionary outcomes are highly sensitive to low probability events which are unlikely to be replicated across the vast majority of alternative histories of life (Powell & Mariscal 2015, 3).

According to the RCT we should, for instance, expect bird dispersal to be unpredictable. This unpredictability is not due to some epistemic lack in us, but because of the complexity and fragility of macroevolutionary outcomes.

Questions of contingency matter for big biological questions. First, if life is contingent, then it won't be governed by general laws (Beatty 1996; Mitchell 1997; Sober 1997). Second, if macroevolutionary outcomes are contingent, then microevolutionary approaches might not be sufficient to understand them, necessitating an autonomous science of macroevolution (Grantham 1999; Turner 2011; McConwell & Currie 2017). Third, life's contingency suggests that our evolution wasn't foreordained; humans aren't so special. Gould, for instance, highlights one apparently insignificant member of the weird and wonderful Cambrian fauna: *Pikaia*. *Pikaia* is reconstructed as a proto-vertebrate, the earliest known. If *Pikaia* and her relatives didn't happen to survive, Gould suggests, then our species couldn't have evolved. Thus, all our great works owe themselves in part to a tiny lancet-like critter.[18] Fourth, contingency matters for exobiology. If we found life on other planets, will we see similar things to on Earth? When looking for life, what signals should we hunt for (McGhee 2011; Powell & Mariscal 2014; Powell 2020)? I'm here concerned with how comparative thinking leads us to conceive of contingency theses.

Distinguish between *contextual* and *general* contingency theses. The former concern some subset of life, the contingency or otherwise of particular outcomes or patterns, such as patterns of gigantism in birds across various islands.

---

[18] Gould's point doesn't turn on whether *Pikaia* really was the ancestor of all vertebrates, just that something like that lineage was.

The latter says something like: macroevolution is itself contingent. I'll argue that very general forms of the thesis are conceptually and empirically intractable (Losos 2018 argues for a similar conclusion). Does anyone advance general versions of the thesis? Yes:

> Although Gould tended to focus on animal morphology, he argued that 'almost every interesting event of life's history falls into the realm of contingency' (p. 290) – remarks which suggest that the RCT was proposed not as a narrow claim about the evolution of animal body plans, but rather as a general thesis about the grand-scale organization of life on the Earth (Powell & Mariscal 2015, 2–3).

Philosophers interested in Gould's discussion have focused on various notions of 'contingency' he appears to conflate. For instance: should we imagine replays from identical starting points or from slightly different starting points? John Beatty distinguished between contingency *per se* – initial conditions are insufficient for an outcome – and contingent *upon* – some initial conditions are necessary for an outcome (Beatty 2006). Gould seems to be suggesting both that *Pikaia's* surviving the Cambrian is necessary for vertebrate radiation and that *Pikiai's* survival was not guaranteed (that is, the conditions of the Cambrian weren't sufficient alone for *Pikia's* ancestors surviving and flourishing). Regarding contingency *per se*, it is tempting to link contingency with indeterminacy. In a deterministic world, conditions at some arbitrary time, and the laws of nature, are in combination sufficient to decide the state of the world at any other time. In such a world, if we started from identical conditions, we'd get identical replays. Gould is not best read as appealing to determinism or indeterminism. Rather, he's interested in whether some relevant set of initial conditions are sufficient to determine the outcome. That is, biological outcomes are contingent *per se* when the relevant biological facts (as opposed to all the facts) were not sufficient to determine the outcome (see Powell 2009; McConwell & Currie 2017).[19]

Changing the timing of replays (or, more carefully, the factors we hold fixed) generates new questions about life's shape. By the Cambrian, say, much about life was set: it is carbon-based, part of a global oxygen-carbon dioxide exchange, using a four code DNA sequence, the major phyla (including the vertebrates) are in play, and so on. We might ask, *given those conditions*, whether the subsequent vertebrate success story was guaranteed. But on earlier reruns different backgrounds will matter. To see this, let's contrast Gould with Simon Conway-Morris (2002). Conway-Morris is typically associated with

---

[19] Beatty 2016 has since doubled-down on indeterminacy, and Ereshefsky & Turner 2019 have responded.

a non-contingent view of life: if *Pikaia* hadn't made it some suitably human-like things would have evolved anyway. However, contingency still looms large. Conway-Morris reckoned life's emergence is very unlikely, its arising in the first place relies upon a very particular set of improbable events. On this view, if we rewound the tape to before life's emergence on Earth, we should bet against life happening at all. So, both Gould and Conway-Morris agree that life is contingent, but disagree on where the contingency is located.

So, by restarting at different points in time, and by holding different backgrounds fixed, a plethora of different claims about contingency are generated. Despite this, Gould and Conway-Morris are often interpreted as providing *general* accounts of life's contingency or otherwise. Although Conway-Morris thinks life's being on earth is contingent, he does think that if life were to arise, it would – inevitability – lead to relevantly human-like beings. Gould disagrees. Notice differences in scope: asking whether life on earth would be the same if we replayed from the Cambrian is a very different question to asking what life on other planets might be like, or on alternative Earths replayed earlier than life's emergence (see Wong 2019's notion of 'modal range'). General contingency theses, then, say that – across these variations – macroevolutionary outcomes are contingent, while general convergence theses say the opposite.

But what evidence could we have of such theses? Powell and Mariscal have an excellent discussion rooted in comparative thinking. I'll build on theirs, ultimately generating a challenge for evidencing general contingency theses. I'll start with their account of contingency, disambiguating a few concepts and providing some theoretical machinery going forwards. I'll then turn to comparative evidence for or against contingency theses: convergences and unique events. I'll then be in a place to be pessimistic about general contingency theses.

3.1 Morphospace and Contingency

Having evolved at least four times, dog skull morphology is relatively common amongst carnivorous marsupials, while cat skull morphology is unheard of. Competing explanations for this uniqueness appeals to divergences in placental and marsupial development. Perhaps the marsupial's extremely early birth, and the requirement to reach the safety of the pouch, necessitates the early development of strong jaws; perhaps the relative simplicity of marsupial tooth development blocks the evolution of catlike dental morphology. This is a contextual contingency question: what does the evolution of catlike morphology depend upon? In this section, I'll connect discussions of evolutionary profiles, such as that of cat skull evolution, with the notion of 'morphospaces' before connecting this with Powell and Mariscal's account of contingency.

In a *theoretical* morphospace, morphology is conceived via a limited set of dimensions which together generate the relevant forms (McGhee 1999; Maclaurin 2003; Currie 2012). The paradigm example is David Raup's pioneering work on shells (Raup 1966). Shell growth can be geometrically modelled. Raup took three dimensions: the translation rate, expansion rate, and the distance of the generating curve from the coiling axis, and from this generated a three-dimensional space of possible shells. The theoretical shell space is the geometric result of those dimensions' unfurling. Onto this theoretically generated space Raup layered an *empirical* morphospace. An empirical morphospace describes the morphology of actually occurring shells in nature. By comparing theoretical and empirical morphospaces (see Figure 10), Raup generated a set of questions. For each part of unoccupied theoretical morphospace we can ask *why hasn't this region been colonized?* For each part of occupied theoretical morphospace we can ask *why have these lineages colonized these morphologies?* And indeed, of each lineage we can ask *why is the lineage distributed across morphospace as it is?*

As George McGhee conceptualizes it, two broad factors explain the distribution of actual forms across a theoretical morphospace (McGhee 2011). First,

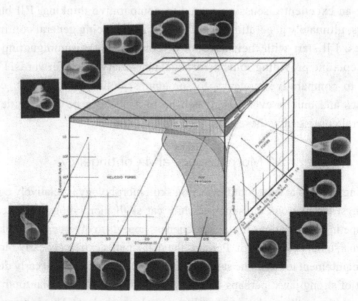

Figure 10 Raup's three-dimensional theoretical shell space, with empirical shell space shaded in (from Raup 1966, 1884 © SEPM Society for Sedimentary Geology.)

developmental possibility. From certain regions, some morphologies might not be 'reachable'. Hypotheses concerning the lack of marsupial cats have this form. Features of marsupial ontogeny block routes to some areas of morphospace – routes that are open to placentals. Second, *functional possibility*. Here, McGhee refers to the fitness of those morphologies. Some morphologies might be bad bets: too inefficient, structurally unsound, carrying major energetic costs, or adding a critical disadvantage in the endless rounds of predation, predation-avoidance, mating, and ensuring the next generation's survival, that are natural selection's levers.

At low dimensions, morphospaces are a useful way of understanding how comparative thinking builds evolutionary profiles. Although morphospaces are focused on morphology, it doesn't take much imagination to extend the idea to other elements of life: a space of possible genes or developmental resources, for instance.

Raup's shell morphospace and carnivore skull space we saw in the introduction are both *contextual* morphospaces in two senses.[20] First, in terms of theoretical morphospace, they capture a particular kind of morphology such as shells or skull shapes. Second, the latter's empirical morphospace is constrained to mammalian carnivores. As such, they are evolutionary profiles useful for probing contextual contingency theses.

McGhee suggests we can also consider life's shape in terms of *total morphospace*. That is, instead of restricting ourselves to a simple range of morphologies, we ask ourselves about the shape of life generally. Given the *whole range* of how living systems might be, why are some areas occupied and others not? Although as we'll see I'm no fan of general morphospaces, it is worth getting clear on what they are and how they relate to contingency theses before turning to complaints. Following the distinction between functional and developmental possibility mentioned earlier, we can conceive of this enormous theoretical morphospace as shaped by these two constraints on biological possibility.

> We now are in the position to consider the spectrum of all possible existent, nonexistent, and impossible biological forms. Let us consider the developmental constraint boundary ... to spatially portray the spectrum of all biological forms that can be developed by life on Earth. That is, forms within this region can be developed by organisms present on Earth, but forms outside this region cannot. We now see that four different regions of potential form exist within the morphospace (McGhee 2012, 250).

---

[20] One difference is in the order of construction: the shell space is constructed *a priori* from Raup's dimensions, while the skull space is abstracted from data via principle component analysis.

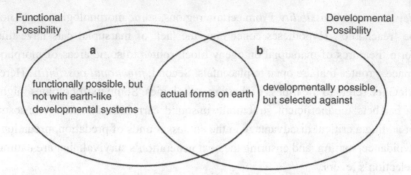

Figure 11 Total Morphospace (from Currie 2012, 586) © Springer Nature

In Figure 11, we can think of the actual forms of life on Earth as occupying the intersection of the functionally and developmentally possible regions of total morphospace.

Following McGhee, we have four spaces to consider: (i) those morphologies that are functionally and developmentally possible, (ii) those which are functionally possible but not on Earth, (iii) those which are developmentally possible but are selected against, and (iv) those which are neither functionally nor developmentally possible. Such a total morphospace gives us conceptual grip on what might be meant by a general contingency thesis. Of the occupied areas of the total morphospace, what proportion is repeatedly occupied, what proportion is occupied once, and what proportion is empty? Further, what determines movement through morphospace: development, selection, chance, or something else? McGhee explicitly links consideration of total morphospaces to contingency.

> At first this question may seem to be so abstract as to be of no real importance, but it is of critical importance in our consideration of the question of predictability in evolution . . . We know that much of the convergent evolution of life on Earth is driven by developmental constraint, yet we do not know the answer to two of the most fundamental questions concerning developmental constraint itself: 'How did development originate [on Earth]? 'and 'How did the developmental repertoire evolve? ' (Müller 2007, 944). Only when we have the answers to those two questions can we make predictions about the possible evolution of alien developmental repertoires, and whether or not those developmental repertoires might be similar to those found in Earth life (McGhee 2012, 251–252).

Or as I would put it: in building an evolutionary profile of all life, we need to understand to what extent life's path on Earth has been constrained and shaped by shared, inherited developmental systems. Again, we can compare these with local contingency theses, such as *the evolution of catlike dental morphology depends on non-marsupial development*. We'd here envision the space of possible marsupial dental morphology, with molar-incisor gaps being functionally possible but not developmentally possible.

Morphospaces help us understand the evolutionary profiles at the integrative core of comparative thinking, as well as understand both contextual and general contingency theses. Let's combine this with Powell & Mariscal's account of radical contingency. Here it is:

> We will consider an outcome *radically contingent* when its existence depends on events occurring during a given evolutionary path that are unlikely to be replicated across the vast majority of alternative evolutionary histories. Radically contingent systems may be contrasted with *robustly convergent* systems in which many outcomes remain accessible from distant evolutionary trajectories. (Powell & Mariscal 2015, 2).

When we ask whether some evolutionary outcome is radically contingent, we're asking, upon surveying total morphospace, how often that outcome occurs. Powell and Mariscal's account focuses on phyla-level or other broad contexts. Even if some outcomes are robust given relatively constrained situations (say, if given a marsupial-like ancestry doglike skulls are likely to evolve), the point is that they are fragile across total morphospace (say, only in the mammal-like histories do we get doglike skulls). Powell and Mariscal's account appears to conflate two distinct notions: probability and dependency, which I'll disambiguate.

Dependency and probability can be pulled apart. The molar-incisor gap might depend on a placental-like developmental system. That is, molar-incisor morphology space is only occupied by, and accessible to, critters with those developmental systems. But this isn't fully informative of the likelihood of molar-incisor morphology. Likelihood is commonly understood as a kind of conditional probability: how probable an observation is given a hypothesis. We can consider other conditional probabilities as well, say, the probability of some outcome given some prior event. It could be that, given our understanding of placental developmental systems, it is highly likely that we'll see catlike teeth. Or, it could be that although placental development is required, it is still unlikely that molar-incisor gaps will evolve. Even if the existence of the incisor-molar gap is highly dependent on having a placental developmental system, it doesn't follow that it will be replicated across alternative evolutionary histories. Moreover, patterns of dependence can be highly complex and interwoven. One way of getting catlike teeth is to have the right developmental system and to have selection for strangulation. But there may be others, and there may be more dependencies. That is, multiple routes might be available through morphospace: the dependencies may be partial. So, we might come to understand the (let's call it) *topography* of morphospace: the patterns of dependence between the dimensions and thus the routes of accessibility through it. But knowing topography doesn't tell us about conditional nor total probability. There might be multiple routes, but each of these might be extremely improbable. So, determining a contingency thesis involves asking three questions: one, what is the space of possibility? That is, determining the relevant dimensions. Two, what is the topography? That is, discovering the dependency relations across those dimensions. Three, what are the conditional probabilities? That is, understanding how probability is distributed across the morphospace.

Contingency theses, then, can be disambiguated into claims about the topography of empirical morphospaces and the distribution of probability across them. An outcome might be highly contingent insofar as there is only one route to it; however, that very same outcome might be highly uncontingent due to being highly probable. Because I'm unsure of how to determine probability distributions across morphospaces, I'll focus on topography going forwards. That is, we'll understand radically contingent outcomes as being inaccessible across morphospace: there are few routes to that location[21] (Inkpen & Turner 2012 take a similar approach). Such contextual morphospaces consist in (1) a theoretical morphospace consisting of a few explicitly characterized dimensions; (2) an empirical morphospace capturing which critters occupy the theoretical space.

So, a contextual contingency thesis says that, given contextual morphospace, some outcome is accessible by very few routes. By contrast a general contingency thesis says, given total morphospace, outcomes are accessible by very few routes. How might we evidence such theses? Comparative thinking provides some answers and reasons to worry about general theses.

3.2 Convergence and Singletons

Contingency theses are questions about the modal robustness of life. They concern not how life actually unfurled, but how it could have. Evidential access to possibility is tricky when you've only one case. In establishing a causal connection scientists run multiple experiments, varying controls and hunting for confounders. You can't say that mummichogs locate water by looking for reflectance patterns just by observing a single fish. But life on Earth has arisen but once and we've not found it elsewhere. Nonetheless, we can consider two sources of empirical evidence (Sterelny 2005, Mariscal 2015). First,

---

[21] Such an account is vulnerable to a recent complaint from Wong: some accounts of contingency do not guarantee the evitability of a contingent outcome. That is, on this account it is possible for an outcome to be contingent, as there is only one route to it in morphospace, but nonetheless guaranteed to occur because the probability of that route is one (Wong 2020).

experiments. We can rerun evolutionary events in controlled environments using small, tractable critters such as *E. coli*. These have been much-discussed, and no doubt provide insights into evolutionary patterns and the robustness of some outcomes (Beatty 2006; Desjardins 2011b; Parke 2014). However, questions of scale loom large here: it is difficult to ascertain the relationship between repeated or non-repeated mutation events in *E. coli* under highly controlled experimental conditions and, say, the contingency or otherwise of *Pikaia* and her relatives' survival. The second source of evidence – and my focus here – involves comparative thinking. Although life has only arisen once, it is marked by patterns of radiations, adaptations, and so forth. These outcomes could act as 'natural experiments' and inform us about contingency.

Convergence, the repeated evolution of similar traits, is the most discussed evidence in this arena (Conway Morris 2002; Sterelny 2005; Currie 2012; Powell 2012). As we've seen, island gigantism occurs over and again across birds – even parrots – and this, you might think, provides evidence of the robustness of that kind of evolutionary event. We can conceive of each *Aves* lineage reaching an island, shedding flight, and becoming enormous, as an experimental run. This enables us to determine various dependencies between those events. For instance, because smaller islands are relatively limited in terms of niche space, we should expect only a single lineage to become gigantic – and this is indeed what we see. Examining patterns of convergence, then, can empirically inform us about the contingency (or lack thereof) of evolutionary outcomes: convergence speaks in favour of evolution's repeatability.

William Wong has argued that in addition to convergence we can examine unique lineages, what I'll call *singletons*.

> These are forms that, for one reason or another, have evolved uniquely. Often peculiar and distinctive, these forms constitute direct counter-examples to the robust view of life in that their evolution has been singular within the corresponding range. But their evidential role extends further than acting as mere counter-examples: the reasons for their singular evolution are informative of the way in which evolution has failed to be robust (Wong 2019, 22).

You might worry about unique singletons: aren't all lineages unique? Wong argues that evolutionary uniqueness can be identified so long as we're clear on what trait and context interest us. As we've seen, cats are unique across *Carnivora* and carnivorous marsupials for their molar-incisor gaps and killing via strangulation; they're potentially not unique across life (for all I know, there may be other critters with the gap or with that killing method) and they're certainly not unique in terms of having, say, a *Carnivora* jaw. Similar could be said concerning convergences. Bats and birds are convergent vis-à-vis *having wings*, but they're not vis-à-vis *wing structure*, as birds' wings are stretched out across a single 'finger', while bat wings are more articulated and use three digits. Wong argues that the occurrence of singletons speak in favour of contingent evolutionary outcomes at least in a local way (see also Losos 2017, chapter 3).

We thus have a strategy for investigating life's shape: survey convergences and singletons. There are some ambitious examples out there. Both Conway-Morris and McGhee have published long lists of convergences across life's scales; and Geerat Vermeij has considered singletons (2006). But what are we to make of such surveys? Jonathon Losos, for one, thinks we should do better:

> We could argue these points back and forth until we're blue in the face. I'd throw out the platypus, you'd counter with convergent hedgehogs; I'd postulate the unique, algae-encrusted upside-down-hanging tree sloth, you'd retort with bipedal-hopping mice independently evolved on three continents. And that is how, essentially, the controversy has been debated historically, by compiling lists and telling stories (Losos 2017, 106).

Losos thinks convergences can be studied more carefully. Examine changes on a small scale, across a particular radiation event, and apply statistical methods to determine the causal specs (see also Bromham 2016). He also, similarly to me, rejects general contingency theses. However, I will sound a note of caution regarding too much emphasis on statistical, experimental methods: we needn't read surveys as mere lists but in terms of comparative thinking. When I appeal to the convergence between, say, hedgehogs and echidna, I'm situating them in terms of ancestry and adaptation; how their particular ancestral and environmental history shaped them into that morphology. Restricting ourselves to statistically amenable events has the same challenge as laboratory experiments: scale. It isn't clear what the relationship is between a single lineage of lizards radiating across several islands and *Pikaia's* survival or otherwise. This doesn't mean that Losos' approach is mistaken, rather his arguments against considering convergences in less 'experimental' contexts are too quick. Insofar as Losos and I both reject uncontextual contingency theses and demand that convergences and singletons be placed in a comparative context, however, we're very much on the same page.

To sum up, some access to the repeatability or otherwise of evolutionary outcomes can be had via examining life across multiple scales and angles: this is how an evolutionary profile reveals itself. We can run lab experiments on *E. coli* and similar critters, we can examine small-scale radiations and bring statistical methods to bare, and we can compile convergences and singletons across life and through the fossil record. Together, this can provide access to the patterns of dependency across local morphospaces, and thus inform us about contextual contingency theses.

But what about general contingency theses?

Concerning total morphospace, not just any old convergence will do. Learning that, say, carnivorous marsupials often converge on a doglike skull tells us about the repeatability of the skull morphology *in marsupials*, not across total morphospace. It doesn't, for instance, tell us what to expect on other planets. After all, we don't know how likely marsupial-like morphology and developmental systems are. One answer to this problem is to identify particular convergences as being particularly telling: it isn't all convergences, but those with a particular set of properties which can speak for or against general contingency theses. The most developed version of this strategy is Powell and Mariscal's. They argue that a set of convergences test general contingency theses when they meet three conditions.

First, the characterization shouldn't be too abstract (they call this 'specificity'). Recall McGhee's category of 'carrion eaters', which included carrion beetles and hyenas. Even if, squinting our eyes, we agree these are ecological equivalents, we might nonetheless note the equivalence is realized via massively divergent structures. Recall the shallowness problem from Section 1: if convergences are only identified via highly abstract categories, that doesn't tell us much about the specific structures that instantiate them, limiting their evidential value. Predicting that, say, life on other planets will likely evolve light detecting apparatus (assuming there's light!), or that there will be predators and prey, seem like safe bets. But this barely contradicts the radical contingency thesis, as this concerns the distribution of structures and mechanisms across total morphospace. So, less abstract convergences, those involving the evolution of structural and molecular elements are more telling. A classic example is the evolution of lens camera eyes (although there are important differences in how lens eyes work in different taxa).

> The image-forming eye includes several distinct eye types – variations on camera and compound configurations – each of which is highly structurally specific, including particular cornea, lens and retina configurations ... . Camera-type eyes, for example, have evolved not only in vertebrates, arthopods, molluscs and cnidarians, but even in microbial eukaryotes (Powell & Mariscal 2015, box 1).

Second, the convergences should be highly *independent*, that is, they should lack shared developmental resources. This is because the occurrence of parallel evolution raises no trouble for contingency theorists: if developmental constraint is powerful, it is no surprise to find repeated evolution. Often, phylogenetic distance is used as a proxy for independence, but this is a proxy only: highly conserved 'deep' homologies could still play a large role in shaping convergent evolution. Ideally, a causal understanding of the actual developmental mechanisms should be had. As Powell and Marsical point out, although there are empirical and conceptual challenges to identifying independence mechanistically, it is nonetheless tractable.

Third, we should consider the *scope*, or range and ubiquity of the conditions under which the convergent traits can obtain. Scope encompasses environments and traits. Evolving a molar-incisor gap requires having mammalian-style dental morphology in the first place; evolving island gigantism requires the existence of islands (or at least climes with island-like isolation). The repeatability of very narrow scopes can turn in part on the commonness of those conditions, thus making them unhelpful evidence regarding the radical contingency thesis. Those obtainable under more general conditions are more telling.

Powell and Mariscal's idea, then, is that convergences that are high in detail (not too abstract), independent, and with broad scope, truly challenge the radical contingency thesis. In other words, paths exhibiting convergences across total morphospace demonstrate how evolution is highly constrained. By their lights, showing that two or more lineages, despite not sharing developmental resources nor particularly similar environments, evolved remarkably similar structures, shows that paths in morphospace are no way near that constrained; or at least are not constrained by contingent biological features. Powell and Mariscal's account is steeped in comparative thinking, envisioning lineages as shaped by ancestry and adaptation. However, the comparative nature of the work exactly makes me pessimistic about general contingency theses, and whether Powell and Mariscal's convergences can bear on them. This challenge will bring us back to the locality of comparative thinking: its power lies in how lineages are shaped by ancestry and adaptation, and as the reach of general contingency theses goes beyond these, they lack both the conceptual and empirical clarity comparative thinking affords.

3.3 Challenging General Contingency Theses

I've suggested that we can understand contingency theses using morphospaces. Contextual contingency theses claim that certain routes through low-dimensional morphospaces or across particular taxonomic groups are highly path dependent. General contingency theses claim that routes through total morphospace are highly path dependent. We've also identified a strategy for empirical traction on contingency theses: examine patterns of contingency and singletons. We're now in a place to worry about total morphospace and thus general contingency theses.

As we've seen, comparative evidence is local: sensitive to description, situated within a phylogeny, and specific to the trait at hand. But radical contingency theses are highly *un*local: they concern the shape of total morphospace. This raises two related problems. First, in total morphospace, there is no theoretical morphospace in which to situate the empirical morphospace. Second, there is no independent reason to think that contingency (or robustness!) in one region or dimension of total morphospace will translate to contingency (or robustness!) in another.

Comparative thinking is sensitive to description. That is, the inferences we can draw, and which comparisons or contrasts are relevant, are highly sensitive to how we characterize our comparisons. *H. inexpectatus* is not unique for being a parrot that weighs more than three kilograms – Kakapo manage that – but they are unique for being the largest parrot (thus far). Similarly, they're not unique *among aves* for being island giants but they are, so far as we know, unique amongst the parrots. So, our scope, how we characterize our target, makes a difference. One way of understanding theoretical morphospaces is as a precise and abstract way of characterizing a trait. In the carnivore skull space, skulls are very specifically described in terms of two dimensions along which skull morphology may be transformed. In Raup's shells, shells are described in terms of three dimensions of growth. This provides a geometric, precise, and specific base for comparisons. It is exactly because we can characterize the theoretical morphospace as we do that we can lay the empirical morphospace over it. This allows comparative thinking to come into its own: identifying puzzling gaps (why no marsupials in cat space), convergences (marsupials with doglike skulls), and generating and testing hypotheses regarding the evolutionary profile.

By contrast, a total morphospace does not specify a trait at a level of specification, it rather asks us to imagine all character states at all levels of specification. As such, there is no theoretical grounding with which to compare the actual distribution of life to. There being a theoretical morphospace relies upon characterizing a trait: and general morphospaces do not do so. Because the traits which the space is based on are unspecified there is no way to, for instance, identify empty locations within it. The power of comparative thinking depends in part on empirically rooted descriptions of characters. A total morphospace is, in a sense at least, characterless, and thus comparative evidence loses its power. As such, it is unclear what to make of appeals to Powell and Mariscal's unabstract, independent and wide-range convergences. Although indeed they are instances where multiple routes have been taken to similar biological structures, without a general trait-space in which to situate them, we cannot build the kind of evolutionary profile a proper understanding of life's shape demands.

The second worry is more direct. Because comparative evidence is specific to the comparator traits, we're not licensed to infer from a particular trait being robust or fragile to traits being generally so. As we've seen, Worthy et al. make various claims about bird dispersal. Critically, although we should expect a single lineage to go giant on a small island, we shouldn't expect a particular lineage to do so. Islands are small and isolated. They thus have limited niche space and cannot support multiple lineages of giant flightless birds. Further, their isolation makes them prone to founder effects: who gets to be the island giant turns on who happens to turn up. Note the specificity: learning that the diversity of a small island's giant birds is predictable doesn't grant us leave to infer that which bird lineages will be giants is predictable. We have robustness on one level – diversity – and fragility on another – phylogeny. I've argued that this is a characteristic constraint on comparative reasoning. Without extra considerations either way, learning that some outcome specified one way is robust or fragile doesn't grant us reason to think that the outcome specified differently is similarly fragile or robust. You might worry that my characterization of specificity takes things far too far. After all, many inferences in biology rely on aspects of different characters being coupled (Finkelman 2019): both phylogenetic and convergent inferences do exactly this. But those inferences have specific, well-established grounding in an evolutionary profile. In those instances, we have good reason to believe them coupled. Without such reasons, the specificity of comparative thinking blocks the inference.

The upshot is learning that, say, the evolution of lens-eyes is a surprisingly robust evolutionary outcome across a wide range of conditions and developmental systems doesn't grant us leave to say that other features (even those associated with lens-eyes) are similarly robust, much less to think that lens-eye-evolution is representative of total morphospace. So, Powell and Mariscal's strategy of picking out a particular set of convergences will not aid us in exploring the shape of total morphospace. Having said this, insofar as they seek to show that there might be some rather general things to be said about the evolution of certain traits within comparative constraints, their argument – if successful – doesn't clash with locality in the sense I've discussed it.

Another strategy is to think of general contingency theses as, as it were, collections or summaries of contextual contingency theses. As we determine robustness or contingency at particular descriptive and phylogenetic grains, we might slowly build up a picture of the fragility of evolutionary outcomes generally. This is fine insofar as we want to identify sets of contextual theses with a general thesis, but often this doesn't capture what folks have in mind when they discuss life's radical contingency (or robustness). As we saw earlier, such theses are about the 'grand scale organization of life'. The shape of life writ large as McGhee tackles it might be outside of our empirical and conceptual remit. Moreover, it isn't obvious to me why we'd be so interested in conjoined contextual theses: given the locality of comparative thinking, I'm inclined to think it is how robust or fragile evolutionary outcomes are when ancestrally and adaptively situated that is interesting and informative. Locality could lead to scepticism about the broad generalities like those Powell and Mariscal tackle as well. Focusing on less ambitious contextual analyses we are led down productive, iterative pathways: consider again the work on cat skulls in the introduction, where consideration of the morphology of jaw musculature lead to a discussion of adaptation, feline ancestry, the relationship between themselves and other mammals, both placental and marsupial, and the construction of various evolutionary profiles. I'm less confident of the fruitfulness of more abstract, general theses (although see Powell 2020 for an impressive attempt).

It is particularly in combination that these two complaints constitute a substantive challenge to general contingency or convergence theses. Because total morphospace doesn't characterize a trait, it cannot be understood as an evolutionary profile. Because overcoming specificity requires situating lineages in an evolutionary profile, singletons or convergences can at best only dimly inform us regarding decontextual contingency theses about total morphospace.

To what extent evolutionary outcomes *qua* evolutionary outcomes are robust or fragile is – according to this challenge anyway – a mistaken question. Where does this leave us? I don't think that abandoning general contingency theses undermines studying life's shape. It instead encourages us to see that work in a far more piece-meal way. Gould may very well be right (or wrong!) that major events in life's history were radically contingent, and that this matters for the role of laws in biology, for the evitability of humans, for the autonomy of macroevolutionary science, and for exobiology, but this needn't be thought of in terms of a thesis about total morphospace. Instead, these are a series of theses about local morphospaces. But remember, 'local' in my sense needn't mean ungrand, or not ranging across much of life. It simply means constrained by comparative thinking in just the way that general contingency theses are not.

Consideration of the shape of life takes us to the very edges of comparative thinking and, perhaps, striving for generality across life might strain it to breaking point.

Conclusion

Living systems evolve. Diversity accumulates as lineages split, mutations arise, and the vagaries of birth and death sum up to extinction and speciation. Simultaneously, stable patterns arise and are maintained through complex interactions of mating events, gene transfer and the culling of variants less suited to their environments. One lineage of *Carnivora*, the felines, struck upon a novel jaw structure which underwrote a novel method for dispatching prey. Their environments proffered opportunity for adaptation and their placental ancestry bequeathed the resources required for evolving to their way of life. Similar stories could be told for all life, and this gives comparative thinking its empirical power.

Comparative thinking confronts the nature of living systems by integrating processes of adaptation, development, shifting environments, and ancestry – the history and capacities of lineages – into evolutionary profiles. Such profiles provide rich resources for understanding and empirically investigating lineages. As in animal cognition, comparative thinking explains why some experiments are designed as they are, and on what basis we might accept or reject hypotheses about animal minds. But comparative thinking also confronts the limitations of what we might know about life: it is a fundamentally local kind of knowledge. These limitations might mean that considering life at its grandest scales requires going beyond comparative thinking, leaving behind its considerable empirical power. Biologists think in multiple ways, appropriately given life's multiplicity, but a powerful and fundamentally biological way of thinking is comparative.

Bibliography

Beatty, J. (2016). What are narratives good for? *Studies in History and Philosophy of Science Part C: Studies in History and Philosophy of Biological and Biomedical Sciences*, 58, 33–40.

Beatty, J. (2006). Replaying life's tape. *The Journal of Philosophy*, 103(7), 336–362.

Beatty, J. (1995). The evolutionary contingency thesis. *Concepts, Theories, and Rationality in the Biological Sciences*, 45, 81.

Ben-Menahem, Y. (1997). Historical contingency. *Ratio*, 10(2), 99–107.

Boyd, N. M. (2018). Evidence Enriched. *Philosophy of Science*, 85(3), 403–421.

Bressman, N. R., Farina, S. C., & Gibb, A. C. (2016). Look before you leap: visual navigation and terrestrial locomotion of the intertidal killifish Fundulus heteroclitus. *Journal of Experimental Zoology Part A: Ecological Genetics and Physiology*, 325(1), 57–64.

Brigandt, I. (2017). Typology and natural kinds in evo-devo. In L. Nuno de la Rosa & G. Müller (eds.), *Evolutionary Developmental Biology*, Springer. https://doi.org/10.1007/978-3-319-33038-9_100–1.

Brigandt, I. (2013). Integration in biology: philosophical perspectives on the dynamics of interdisciplinarity. *Studies in History and Philosophy of Science Part C: Studies in History and Philosophy of Biological and Biomedical Sciences Volume 44, Issue 4, Part A*, December 2013, Pages 461–465

Brigandt, I. (2010). Beyond reduction and pluralism: toward an epistemology of explanatory integration in biology. *Erkenntnis*, 73(3), 295–311.

Brigandt, I. (2007). Typology now: homology and developmental constraints explain evolvability. *Biology & Philosophy*, 22(5), 709–725.

Brigandt, I. (2003). Homology in comparative, molecular, and evolutionary developmental biology: the radiation of a concept. *Journal of Experimental Zoology Part B: Molecular and Developmental Evolution*, 299(1), 9–17.

Brigandt, I. & Griffiths, P. E. (2007). The importance of homology for biology and philosophy. *Biology & Philosophy*, 22(5), 633–641.

Bromham, L. (2016). Testing hypotheses in macroevolution. *Studies in History and Philosophy of Science Part A*, 55, 47–59.

Brown, C., Laland, K., & Krause, J. (eds.). (2008). *Fish Cognition and Behavior*. John Wiley & Sons.

Brown, R. L. (2014). What evolvability really is. *The British Journal for the Philosophy of Science*, 65(3), 549–572.

Buckner, C. (2013). Morgan's Canon, meet Hume's Dictum: avoiding anthro-pofabulation in cross-species comparisons. *Biology & Philosophy*, 28(5), 853–871.

Calcott, B. (2009). Lineage explanations: explaining how biological mechanisms change. *The British Journal for the Philosophy of Science*, 60(1), 51–78.

Calcott, B., Levy, A., Siegal, M. L., Soyer, O. S., & Wagner, A. (2015). Engineering and biology: counsel for a continued relationship. *Biological Theory*, 10(1), 50–59.

Candea, M. (2018). Comparison in Anthropology: the improbable method. Cambridge University Press, Cambridge.

Currie, A. (2019a). *Scientific Knowledge and the Deep Past: History Matters*. Cambridge University Press.

Currie, A. (2019b). Simplicity, one-shot hypotheses and paleobiological explanation. *History and Philosophy of the Life Sciences*, 41(1), 10.

Currie, A. M. (2019c). Epistemic optimism, speculation, and the historical sciences. *PTPBio*, 11(7), 2011–2015.

Currie, A. M. (2018). *Rock, Bone, and Ruin: An Optimist's Guide to the Historical Sciences*. MIT Press.

Currie, A. (2016). Ethnographic analogy, the comparative method, and archaeological special pleading. *Studies in History and Philosophy of Science Part A*, 55, 84–94.

Currie, A. M. (2014). Venomous dinosaurs and rear-fanged snakes: homology and homoplasy characterized. *Erkenntnis*, 79(3), 701–727.

Currie, A. (2013). Convergence as evidence. *The British Journal for the Philosophy of Science*, 64(4), 763–786.

Currie, A. M. (2012). Convergence, contingency & morphospace. *Biology & Philosophy*, 4(27), 583–593.

Currie, A. & Killin, A. (2016). Musical pluralism and the science of music. *European Journal for Philosophy of Science*, 6(1), 9–30.

Currie, A. & Walsh, K. (2018). Newton on Islandworld: Ontic-driven explanations of scientific method. *Perspectives on Science*, 26(1), 119–156.

Dacey, M. (2016). The varieties of parsimony in psychology. *Mind & Language*, 31(4), 414–437.

Desjardin, E. (2011). Historicity and experimental evolution. *Biology & Philosophy*, 26, 339–364.

Dunbar, R. I. (2009). The social brain hypothesis and its implications for social evolution. *Annals of Human Biology*, 36(5), 562–572.

Dwyer, D. M. & Burgess, K. V. (2011). Rational accounts of animal behaviour? Lessons from C. Lloyd Morgan's canon. *International Journal of Comparative Psychology*, 24(4), 349–364.

Emmerton, J. (2001). *Birds' Judgments of Number and Quantity*. Avian Visual Cognition.

Ereshefsky, M. (2012). Homology thinking. *Biology & Philosophy*, 27(3), 381–400.

Ereshefsky, M. (2009). Homology: integrating phylogeny and development. *Biological Theory*, 4(3), 225–229.

Ereshefsky, M. (2007). Psychological categories as homologies: lessons from ethology. *Biology & Philosophy*, 22(5), 659–674.

Ereshefsky, M. (1998). Species pluralism and anti-realism. *Philosophy of Science*, 65(1), 103–120.

Ereshefsky, M. & Turner, D. (2019). Historicity and explanation. *Studies in History and Philosophy of Science Part A*, 65, 103–120.

Finkelman, L. (2019). Betting & hierarchy in paleontology. *Philosophy, Theory, and Practice in Biology*, 11, 1–6.

Fitzpatrick, S. (2017). Against Morgan's Canon. In K. Andrews & J. Beck (eds.), *The Routledge Handbook of Philosophy of Animal Minds*, Routledge NY, pp. 437–447.

Fitzpatrick, S. (2008). Doing away with Morgan's Canon. *Mind & Language*, 23(2), 224–246.

Forber, P. (2009). Introduction: A primer on adaptationism. *Biology & Philosophy*, 24(2), 155–159.

Ghiselin, M. T. (2016). Homology, convergence and parallelism. *Philosophical Transactions of the Royal Society B: Biological Sciences*, 371(1685), 20150035.

Ghiselin, M. T. (2005). Homology as a relation of correspondence between parts of individuals. *Theory in Biosciences*, 124(2), 91–103.

Godfrey-Smith, P. (2001). Three kinds of adaptationism. In S. H. Orzack & E. Sober (eds.), *Adaptationism and Optimality*, Cambridge University Press, pp.335–357.

Goswami, A., Milne, N., & Wroe, S. (2011). Biting through constraints: cranial morphology, disparity and convergence across living and fossil carnivorous mammals. *Proceedings of the Royal Society B: Biological Sciences*, 278 (1713), 1831–1839.

Goswami, A., Weisbecker, V., & Sánchez-Villagra, M. R. (2009). Developmental modularity and the marsupial–placental dichotomy. *Journal of Experimental Zoology Part B: Molecular and Developmental Evolution*, 312(3), 186–195.

Gould, S. J. (1990). *Wonderful Life: The Burgess Shale and the Nature of History*. WW Norton & Company.

Gould, S. J. & Lewontin, R. C. (1979). The spandrels of San Marco and the Panglossian paradigm: a critique of the adaptationist programme. *Proceedings of the Royal Society of London. Series B. Biological Sciences*, 205(1161), 581–598.

Grantham, T. A. (2007). Is macroevolution more than successive rounds of microevolution? *Palaeontology*, 50(1), 75–85.

Grantham, T. A. (1999). Explanatory pluralism in paleobiology. *Philosophy of Science*, 66, S223–S236.

Griffiths, P. E. (2007). The phenomena of homology. *Biology & Philosophy*, 22(5), 643–658.

Griffiths, P. E. (1996). The historical turn in the study of adaptation. *The British Journal for the Philosophy of Science*, 47(4), 511–532.

Guilford, J. P. (1967). *The Nature of Human Intelligence*. McGraw-Hill.

Hall, B. K. (2012). Parallelism, deep homology, and evo-devo. *Evolution & Development*, 14, 33–39.

Hall, B. K. (2007). Homoplasy and homology: Dichotomy or continuum? *Journal of Human Evolution*, 52(5), 473–479.

Hall, B. K. (2003). Descent with modification: the unity underlying homology and homoplasy as seen through an analysis of development and evolution. *Biological Reviews of the Cambridge Philosophical Society*, 78(3), 409–433.

Harvey, P. H. & Purvis, A. (1991). Comparative methods for explaining adaptations. *Nature*, 351(6328), 619–624.

Harvey, P. H. & Pagel, M. D. (1991). *The Comparative Method in Evolutionary Biology* (Vol. 239). Oxford University Press.

Haslanger, S. (2000). Gender and race: (What) are they? (What) do we want them to be? *Noûs*, 34(1), 31–55.

Inkpen, R. & Turner, D. (2012). The topography of historical contingency. *Journal of the Philosophy of History*, 6(1), 1–19.

Karin-D'Arcy, M. (2005). The modern role of Morgan's canon in comparative psychology. *International Journal of Comparative Psychology*, 18(3), 179–201.

Kendig, C. (2015). Homologizing as kinding. In C. Kendig (ed.), *Natural Kinds and Classification in Scientific Practice*, (Routledge, pp. 126–146.

Kitcher, P. (2003). *Science, Truth, and Democracy*. Oxford University Press.

Lever, J., Krzywinski, M., & Altman, N. S. (2017). Points of significance: principal component analysis. *Nature Methods*, 14(7), 641–642.

Levy, A. & Currie, A. (2015). Model organisms are not (theoretical) models. *The British Journal for the Philosophy of Science*, 66(2), 327–348.

Lewens, T. (2009). Seven types of adaptationism. *Biology & Philosophy*, 24(2), 161.

Mayr, E. (2000). Darwin's influence on modern thought. *Scientific American*, 283(1), 78–83.

Livezey, B. C. (1992). Morphological corollaries and ecological implications of flightlessness in the kakapo (Psittaciformes: Strigops habroptilus). *Journal of Morphology*, 213(1), 105–145.

Logan, C. J., Avin, S., Boogert, N., Buskell, A., Cross, F. R., Currie, A., ... & Shigeno, S. (2018). Beyond brain size: uncovering the neural correlates of behavioral and cognitive specialization. Comparative Cognition & Behavior Reviews.

Losos, J. (2017). *Improbable Destinies: How Predictable is Evolution?*. Penguin UK.

Love, A. C. (2007). Functional homology and homology of function: Biological concepts and philosophical consequences. *Biology & Philosophy*, 22(5), 691–708.

Love, A. C. (2003). Evolutionary morphology, innovation, and the synthesis of evolutionary and developmental biology. *Biology and Philosophy*, 18(2), 309–345.

Maclaurin J (2003) The good, the bad and the impossible: a critical notice of 'theoretical morphology: the concept and its applications' by George McGhee. *Biology & Philosophy*, 18, 463–476

Mariscal, C. (2015). Universal biology: Assessing universality from a single example. The impact of discovering life beyond earth.

Mayr, E. (1959). Darwin and the evolutionary theory in biology. In B. J. Meggers (ed.), *Evolution and Anthropology: A Centennial Appraisal*, New York: Theo Gaus' Sons, Inc, pp. 1–8.

McConwell, A. K. & Currie, A. (2017). Gouldian arguments and the sources of contingency. *Biology & Philosophy*, 32(2), 243–261.

McConwell, A. K. (2017). Contingency and individuality: A plurality of evolutionary individuality types. *Philosophy of Science*, 84(5), 1104–1116.

Mast, S. O. (1915). The behavior of fundulus, with especial reference to overland escape from tide-pools and locomotion on land. *Journal of Animal Behavior*, 5(5), 341.

McGhee, G. R. (2011). *Convergent Evolution: Limited Forms Most Beautiful*. MIT Press.

McGhee G. R. (1999). *Theoretical Morphology: The Concept and Its Applications*. Columbia University Press, New York

Meketa, I. (2014). A critique of the principle of cognitive simplicity in comparative cognition. *Biology & Philosophy*, 29(5), 731–745

Millstein, R. L. (2000). Chance and macroevolution. *Philosophy of Science*, 67(4), 603–624.

Mitchell, S. D. (1997). Pragmatic laws. *Philosophy of Science*, 64, S468–S479.

Morgan, C. Lloyd. (1894). *An Introduction to Comparative Psychology*. London: Walter Scott.

Morris, S. C. (2003). *Life's Solution: Inevitable Humans in a Lonely Universe*. Cambridge University Press.

Müller, G. B. (2007). Evo–devo: extending the evolutionary synthesis. *Nature Reviews Genetics*, 8(12), 943–949.

Nolan, D (1997). Quantitative parsimony. *Quantitative Parsimony, British Journal for the Philosophy of Science*, 48, 329–343.

Nyrup, R. (2018). *Of Water Drops and Atomic Nuclei: Analogies and Pursuit Worthiness in Science*. The British Journal for the Philosophy of Science.

O'Malley, M. (2014). *Philosophy of Microbiology*. Cambridge University Press.

Orzack, S. H. & Sober, E. (1994). Optimality models and the test of adaptationism. *The American Naturalist*, 143(3), 361–380.

Owen R. (1843). *Lectures on the Comparative Anatomy and Physiology of the Invertebrate Animals*. London: Longman, Brown, Green, and Longmans.

Panchen, A. L. (1999, May). Homology-history of a concept. In G. Bock & G. Cardew (eds.), *Novartis Foundation Symposium*, Wiley, pp. 5–17.Wiley.

Parke, E. C. (2014). Experiments, simulations, and epistemic privilege. *Philosophy of Science*, 81(4), 516–536.

Pearce, T. (2011). Evolution and constraints on variation: Variant specification and range of assessment. *Philosophy of Science*, (78), 739–751.

Pearce, T. (2012). Convergence and parallelism in evolution: A Neo-Gouldian account. *The British Journal for the Philosophy of Science*, 63, 429–448

Pigliucci, M. & Müller, G. B. (2010). *Elements of an Extended Evolutionary Synthesis. Evolution: The Extended Synthesis*. MIT Press.

Potochnik, A. (2010). Explanatory independence and epistemic interdependence: A case study of the optimality approach. *The British Journal for the Philosophy of Science*, 61(1), 213–233.

Powell, R. & Mariscal, C. (2014). There is grandeur in this view of life: the bio-philosophical implications of convergent evolution. *Acta Biotheoretica*, 62(1), 115–121.

Powell, R. & Mariscal, C. (2015). Convergent evolution as natural experiment: the tape of life reconsidered. *Interface Focus*, 5(6), 20150040.

Powell, R. (2007). Is convergence more than an analogy? Homoplasy and its implications for macroevolutionary predictability. *Biology and Philosophy*, 22(4), 565–578.

Powell, R. (2009). Contingency and convergence in macroevolution: a reply to John Beatty. *The Journal of Philosophy*, 106(7), 390–403.

Powell, R. (2020). *Contingency and Convergence: Toward a Cosmic Biology of Body and Mind* (Vol. 25). MIT Press.

Ramsey, G. & Peterson, A. (2012). Sameness in biology. *Philosophy of Science*, 79(2), 255–275.

Raup, D. M. (1966). Geometric analysis of shell coiling: general problems. *Journal of Paleontology*, 40(5), 1178–1190.

Rieppel, O. (2005). Modules, kinds, and homology. *Journal of Experimental Zoology Part B: Molecular and Developmental Evolution*, 304(1), 18–27.

Sanford, G. M., Lutterschmidt, W. I., & Hutchison, V. H. (2002). The comparative method revisited. *BioScience*, 52(9), 830–836.

Sansom, R. (2003). Constraining the adaptationism debate. *Biology and Philosophy*, 18(4), 493–512.

Sayers, K., Lovejoy, C. O., Emery, N. J., Clayton, N. S., Hunt, K. D., Laland, K. N., ... & Strier, K. B. (2008). The chimpanzee has no clothes: a critical examination of Pan troglodytes in models of human evolution. *Current Anthropology*, 49(1), 87–114.

Sober, E. (November 2009). Parsimony arguments in science and philosophy – A test case for naturalism p. Proceedings and Addresses of the American Philosophical Association Vol. 83, No. 2 (pp. 117–155). American Philosophical Association.

Sober, E. (1991). *Reconstructing the Past: Parsimony, Evolution, and Inference*. MIT Press.

Sober, E. (1997). Two outbreaks of lawlessness in recent philosophy of biology. *Philosophy of Science*, 64, S458–S467.

Sober, E. (2005). Comparative psychology meets evolutionary biology. In L. Daston & G. Mitman (eds.), *Thinking with Animals: New Perspectives on Anthropomorphism*, Columbia University Press, pp. 85–99.

Sterelny, K. (2005). Another view of life. *Studies in History and Philosophy of Biol & Biomed Sci*, 3(36), 585–593.

Suzuki, D. G. & Tanaka, S. (2017). A phenomenological and dynamic view of homology: homologs as persistently reproducible modules. *Biological Theory*, 12(3), 169–180.

Tucker, A. (1998). Unique events: The underdetermination of explanation. *Erkenntnis*, 48(1), 61–83.

Turner, A. (1997). *Big Cats and their Fossil Relatives*. Columbia University Press

Turner, D. & Havstad, J. C., 'Philosophy of Macroevolution', The Stanford Encyclopedia of Philosophy (Summer 2019 Edition), Edward N. Zalta (ed.), URL = https://plato.stanford.edu/archives/sum2019/entries/macroevolution/.

Turner, D. (2011). *Paleontology: A Philosophical Introduction*. Cambridge University Press.

Turner, D. (2000). The functions of fossils: inference and explanation in functional morphology. *Studies in History and Philosophy of Science Part C: Studies in History and Philosophy of Biological and Biomedical Sciences*, 31(1), 193–212.

Uyeda, J. C., Zenil-Ferguson, R., & Pennell, M. W. (2018). Rethinking phylogenetic comparative methods. *Systematic Biology*, 67(6), 1091–1109.

Van Valkenburgh, B. & Jenkins, I. (2002). Evolutionary patterns in the history of Permo-Triassic and Cenozoic synapsid predators. *The Paleontological Society Papers*, 8, 267–288.

Vaesen, K. (2014). Chimpocentrism and reconstructions of human evolution (a timely reminder). *Studies in History and Philosophy of Science Part C: Studies in History and Philosophy of Biological and Biomedical Sciences*, 45, 12–21.

Vermeij, G. J. (2006). Historical contingency and the purported uniqueness of evolutionary innovations. *Proceedings of the National Academy of Sciences*, 103(6), 1804–1809.

Wagner, G. P. (2018). *Homology, Genes, and Evolutionary Innovation*. Princeton University Press.

Wagner, G. P. (2016). What is 'homology thinking' and what is it for?. *Journal of Experimental Zoology Part B: Molecular and Developmental Evolution*, 326(1), 3–8.

Waters, C. K. (2007). Causes that make a difference. *The Journal of Philosophy*, 104(11), 551–579.

Witteveen, J. (2018). Typological thinking: then and now. *Journal of Experimental Zoology Part B: Molecular and Developmental Evolution*, 330(3), 123–131.

Witteveen, J. (2016). 'A temporary oversimplification': Mayr, Simpson, Dobzhansky, and the origins of the typology/population dichotomy (part 2 of 2). *Studies in History and Philosophy of Science Part C: Studies in History and Philosophy of Biological and Biomedical Sciences*, 57, 96–105.

Wong, T. W. (2020). Evolutionary contingency as non-trivial objective probability: Biological evitability and evolutionary trajectories. Studies in History and Philosophy of Science Part C: Studies in History and Philosophy of Biological and Biomedical Sciences, 101246. www.sciencedirect.com/science/article/pii/S1369848619300640

Wong, T. W. (2019). The evolutionary contingency thesis and evolutionary idiosyncrasies. *Biology & Philosophy*, 34(2), 22.

Worthy, T. H., Hand, S. J., Archer, M., Scofield, R. P., & De Pietri, V. L. (2019). Evidence for a giant parrot from the Early Miocene of New Zealand. *Biology letters*, 15(8), 20190467.

Worthy, T. H., De Pietri, V. L., & Scofield, R. P. (2017). Recent advances in avian palaeobiology in New Zealand with implications for understanding New Zealand's geological, climatic and evolutionary histories. *New Zealand Journal of Zoology*, 44(3), 177–211.

Worthy, T. H., Tennyson, A. J., & Scofield, R. P. (2011). An early Miocene diversity of parrots (Aves, Strigopidae, Nestorinae) from New Zealand. *Journal of Vertebrate Paleontology*, 31(5), 1102–1116.

Zentall, T. R. (2018). Morgan's Canon: Is it still a useful rule of thumb?. *Ethology*, 124(7), 449–457.

Acknowledgements

I'm grateful to Marta Halina, Sabina Leonelli, Alison McConwell, Aaron Novick, Trevor Pearce, Russell Powell, William Wong, and two anonymous referees for extremely useful and kind feedback on draft material. Also thanks to Grant Ramsey and Michael Ruse. Ideas from this element were presented to Exeter's *Cognition and Culture* reading group, at *Philosophy of Biology at Dolphin Beach 13* and at the *Recent Trends in Philosophy of Biology* conference in Bilkent; thanks to the audiences there. Many thanks to Kimberly Brumble for the wonderful illustrations. Some of the research for this Element was funded by the Templeton World Charity Foundation.

Dedicated to the incomparable Kate.

Elements in the Philosophy of Biology

Grant Ramsey
KU Leuven

Grant Ramsey is a BOFZAP research professor at the Institute of Philosophy, KU Leuven, Belgium. His work centres on philosophical problems at the foundation of evolutionary biology. He has been awarded the Popper Prize twice for his work in this area. He also publishes in the philosophy of animal behaviour, human nature, and the moral emotions. He runs the Ramsey Lab (theramseylab.org), a highly collaborative research group focused on issues in the philosophy of the life sciences.

Michael Ruse
Florida State University

Michael Ruse is the Lucyle T. Werkmeister Professor of Philosophy and the Director of the Program in the History and Philosophy of Science at Florida State University. He is professor emeritus at the University of Guelph, in Ontario, Canada. He is a former Guggenheim fellow and Gifford lecturer. He is the author or editor of over sixty books, most recently *Darwinism as Religion: What Literature Tells Us about Evolution; On Purpose; The Problem of War: Darwinism, Christianity, and Their Battle to Understand Human Conflict;* and *A Meaning to Life*.

About the Series

This Cambridge Elements series provides concise and structured introductions to all of the central topics in the philosophy of biology. Contributors to the series are cutting-edge researchers who offer balanced, comprehensive coverage of multiple perspectives, while also developing new ideas and arguments from a unique viewpoint.

Cambridge Elements ≡

Philosophy of Biology

Elements in the Series

A full series listing is available at www.cambridge.org/EPBY

Printed in the United States
By Bookmasters